Socrates Tenured

Collective Studies in Knowledge and Society

Series Editor

James H. Collier is Associate Professor of Science and Technology in Society at Virginia Tech.

This is an interdisciplinary series published in collaboration with the Social Epistemology Review and Reply Collective. It addresses questions arising from understanding knowledge as constituted by, and constitutive of, existing, dynamic and governable social relations.

Socrates Tenured

The Institutions of Twenty-First-Century Philosophy

Robert Frodeman and Adam Briggle

ROWMAN &
LITTLEFIELD
INTERNATIONAL

London • New York

Published by Rowman & Littlefield International Ltd
Unit A, Whitacre Mews, 26-34 Stannary Street, London SE11 4AB
www.rowmaninternational.com

Rowman & Littlefield International Ltd.is an affiliate of Rowman & Littlefield
4501 Forbes Boulevard, Suite 200, Lanham, Maryland 20706, USA
With additional offices in Boulder, New York, Toronto (Canada), and Plymouth (UK)
www.rowman.com

British Library Cataloguing in Publication Data
A catalogue record for this book is available from the British Library

ISBN: HB 978-1-7834-8309-9
        PB 978-1-7834-8310-5

Library of Congress Cataloging-in-Publication Data Available
ISBN 978-1-7834-8309-9 (cloth : alk. paper)
ISBN 978-1-7834-8310-5 (pbk. : alk. paper)
ISBN 978-1-7834-8311-2 (electronic)

♾™ The paper used in this publication meets the minimum requirements of American
National Standard for Information Sciences—Permanence of Paper for Printed Library
Materials, ANSI/NISO Z39.48-1992.

Printed in the United States of America

Contents

Acknowledgements

We'd like to acknowledge the generous support of the National Science Foundation (NSF). Two grants have contributed to the research in this book:

- Award #1353796, A Study of How Philosophy Research and Practice Can Better Contribute to STEM Research and to Public Policy.
- Award #1445121, EAGER: Research on the Broader Impacts of Basic Science: Gauging the State of the Art.

We also express our thanks to Kelli Barr, who has helped us with both grants, and provided a great deal of the data behind our empirical analysis.

This work forms part of a continuing conversation among friends and colleagues. Without holding them responsible for any of the following, we would like to thank: Evelyn Brister, Keith Wayne Brown, Steve Fuller, Britt Holbrook, Eunice Nicholson, Michael O'Rourke, and David Budtz Pedersen.

Foreword

Steve Fuller

Philosophy has never been comfortable with its status as a discipline in the academy. Even today, the philosophers who most students read in their courses were non-academics: Plato, Descartes, Hume, Mill. In fact, Kant had to dragoon philosophy into the academy to stop the squabbling of the doctors of theology, medicine, and law, which was threatening to tear asunder the late eighteenth-century university. His famous late essay, *The Contest of the Faculties* (1798), set the stage in the nineteenth and early twentieth centuries for philosophy to define the terms on which all knowledge claims might be transacted. The philosopher would be, in effect, the intellectual underwriter for the academy.

Admittedly, the exact nature of this role underwent considerable change from the time of Kant's immediate followers, the German idealists, to its final robust incarnation in the logical positivist movement. What they all shared, however, was a commitment to the philosopher as someone who in his or her own person resolved the contradictory claims made by various academicians. Goethe was often cited as the exemplar who, in this sense, 'lived philosophically'. Of course, academic philosophers continue to pay tribute to this lineage in fields called 'epistemology' and 'philosophy of science'. However, today's versions are paler, more inward-looking versions of what had animated Kant. Indeed, Kant would be disappointed by the extent to which the institutionally dominant 'analytic' school of philosophy appears to have been captured by the very academicians who he would have philosophers domesticate, if not dominate.

Socrates Tenured is most naturally read as mounting a wholesale attack on this decadent state of academic philosophy. But a deeper reading of the book suggests that its authors, the environmental philosophers Robert Frodeman and Adam Briggle, may object to Kant's very tasking of philosophy with the harmonization of academic knowledge production. After all, our authors self-identify as 'field philosophers' (as in 'field researchers'). For them, philosophy's big sociological mistake was to become captive to the emerging 'research university' in the late nineteenth century, whereby specialized doctoral-level training in philosophy became the benchmark of philosophical competence. This started the drive towards inwardness, which has resulted in contemporary academic philosophy being so difficult to fathom, even by interested and otherwise well-informed non-professionals. Frodeman and Briggle go still further: for them this inward turn has spelt the end of philosophy as virtuous inquiry – and the philosopher as someone who by his or her own practice lives some version of a life worth living.

When Frodeman and Briggle debuted this argument in January 2016 in *The Stone*, The *New York Times*' philosophy column, the noted historian of analytic philosophy, Scott Soames, responded by reasserting that philosophy's natural home is its current one, the academy. Soames's view is widely shared by professional philosophers. Unsurprisingly, perhaps, Soames argued from his own expertise, namely, the study of language, logic, and science within analytic philosophy. That Soames's glaring sampling bias has been so easily granted by his philosophical colleagues unwittingly supports Frodeman and Briggle's point about the inward-looking nature of the field.

A striking move in Soames's argument is his quick distinction between the theoretical side of philosophy (i.e. the concern with language, logic, and science) and the practical side (i.e. the concern with goodness, justice, and virtue) – on the basis of which he judges the former to have been always more important than the latter to the development of philosophy. This move is a signature feature of just the sort of institutionalization that Frodeman and Briggle decry. Call it the "Neo-Kantian settlement," after how Kant's followers increasingly thought about philosophy as an academic subject in the nineteenth century. The basic 'Neo-Kantian' idea is that philosophers cannot speak sensibly about the ends of things (i.e. the practical side), but they can say much of value about what follows from presuming certain ends and perhaps even what might facilitate those ends (i.e. theoretical side).

Soames's defence of academic philosophy proceeds by showing the many ways – mainly in the twentieth century – in which philosophy and the special sciences have worked in a mutually enriching fashion. This is supposed to show that far from indulging in academic navel-gazing, philosophers have been both receptive and contributory to the work of other disciplines. Yet, in all the cases cited by Soames, philosophy is operating in what John Locke (cited by Soames approvingly) called an 'underlabouring' capacity: The other disciplines provide the ends, and philosophy then supplies or refines the means. Such is the Neo-Kantian settlement in action. Even when Soames cites philosophers who revolutionized other disciplines, his examples are limited to the philosopher-mathematicians Gottlob Frege, Bertrand Russell, and Kurt Gödel, as if to suggest that philosophy had some natural affinity with pure thought, detached from human ends, which might make it especially well-suited to the mathematical disciplines.

In counterpoint, consider three other philosopher-mathematicians of roughly the same vintage of Frege, Russell, and Gödel: Charles Sanders Peirce, Edmund Husserl, and Henri Bergson. They too engaged in revolutionary thought of at least equal philosophical import – but their effects have been distributed more widely, both in and out of the academy. Whereas Frege, Russell, and Gödel are respected in academic philosophy largely for quite specific and technical contributions which show what philosophers can do to push forward the frontiers of knowledge, Peirce, Husserl, and Bergson made broader contributions of more general human significance which show the sort of people that philosophers invite us to be. Their philosophical projects of, respectively, pragmatism, phenomenology, and vitalism remain part of the cultural landscape today, even if not housed specifically in philosophy departments.

Indeed, contrary to the tenor of Soames's entire argument, philosophy tends to be culturally creative – and socially destabilizing – precisely because it naturally roams over the whole of the human condition, blurring the theoretical–practical distinction. This is the most straightforward way of interpreting the version of Socrates presented in Plato's dialogues. In contrast, the Neo-Kantian settlement would have philosophy lead from its theoretical side, thereby avoiding the moral, political, and religious difficulties which regularly plagued previous philosophers. It was, thus, fit for a privileged place in the academy. This was a familiar move from the seventeenth-century founding of the Royal Society in London, which justified the 'separate yet superior' stance of the emerging sciences of nature. Max Weber even provided a charitable reading of this strategy in the early twentieth century: Academic neutrality on the ends of knowledge in the classroom creates a space that enables students to decide for themselves which knowledge claims are worth pursuing – and to what extent.

A good way to see the effect of Neo-Kantian settlement is in terms of the changing face of Socrates, for whom no philosopher nowadays has a bad word. However, it was only with philosophy's academic domestication that the image of Socrates came to be seen as unequivocally positive. Indeed, it marked the domestication of the image of Socrates himself. Even in the Enlightenment, when one might have expected a uniformly receptive audience for Socrates, Voltaire and Rousseau were divided on his legacy – the former regarding Socrates much more favourably than the latter. However, today's academic Socrates is all about his technical prowess in argumentation rather than the seemingly mischievous and bloody-minded character of his interventions, something which had bothered Rousseau and many others. Today's sanitized Socrates became possible once academics started reading Plato's dialogues quite literally and ignored the performative character of Socratic utterances – let alone any of the larger, often negative, motivational interpretations.

In effect, the academically domesticated Socrates is a dramatic vehicle – a puppet – crafted to defend Plato's doctrines in the face of real and imagined critics. To be sure, the man named 'Socrates' probably existed, but his combination of memorable personality and lack of written trace made him the ideal mouthpiece for Plato, one of the later followers of Socrates. Those who knew Socrates longer seem to have had a less exalted image of him. To continue in this vein is to go down the route pursued by most Plato scholars today, which involves sorting out the Plato-Socrates relationship with an eye to legitimizing current academic philosophical practice. But of course, one might cut to the chase and simply decide whether or not to emulate the Socrates that appears on Plato's pages, regardless of whether that figure existed. Here is where the thesis of *Socrates Tenured* raises some interesting questions.

Socrates Tenured presents two contrary images of what it would mean to be Socrates in today's academic environment. The two images hark back to two schools of Greek thought contemporary with Socrates, each of which has left a linguistic and cultural trace that endures to this day: *sophist* and *cynic*. Frodeman and Briggle's 'field philosopher' struggles between following the way of the sophist and the cynic. While both ways are normally understood in an ironic light, it would be truer to the original Greek context to take both positions straight – which is exactly how the reader should understand Frodeman and Briggle. On the one hand, the sophist believes that philosophy earns its keep only if it can empower the non-philosophical world and will do everything he or she can to aid in that project. From this standpoint, Socrates is a sophist with a scrupulous sense of quality control. On the other hand, the cynic believes that philosophy earns its keep only if it can disabuse people of beliefs that inhibit their ability to live a peaceful existence. From this standpoint, Socrates is a cynic with a good bedside manner, as in the later Wittgenstein's 'therapeutic' approach to philosophy. Readers are invited to consider which proportions work best for them as they try to lead the Socratic life in today's world.

An interesting data point in this discussion is the career of Bertrand Russell, who Soames invokes in criticizing Frodeman and Briggle. Russell was someone who did not merely do philosophy well but also lived a philosophical life. This more practically oriented aspect of Russell's philosophical career – most of it, as measured in years and pages – is typically passed over in relative silence by academic philosophers. Yet, it is precisely the sort of philosophical agency exemplified by Russell – his philosophical personality – which marks him as a genuine 'philosopher' in Frodeman and Briggle's sense. Russell was comfortable making the same argument in the public square as in the seminar room. And I mean here more than simply Russell's famed ability to translate complex ideas for popular audiences. More importantly, Russell's moral stature was evidenced in his willingness to speak his mind on a wide range of issues – with "insulting clarity" to people not inclined to believe him – and accept the consequences, including jail time. In this respect, Russell was very much like the academically unexpurgated Socrates.

Academic philosophers take too much pride in being able to discuss matters in seminar rooms that would cause riots if taken seriously in public. They regard the public very much as Plato did – namely, as mentally unprepared to think deeply and broadly about matters of existential import. But the cost of requiring such 'mental preparation' is that academic philosophers end up with a semi-detached, gamesman-like attitude towards the vital matters with which they are putatively concerned. The relevant sense of 'mental preparation' amounts to attaching the warning, 'Don't try this at home!' to any radical argument or thought experiment which upends our intuitions about reality. This is not good enough for Frodeman and Briggle, who would have philosophers 'walk the walk' as well as 'talk the talk'. To be sure, this injunction still involves a strong sense of 'discipline' but it is one that is perhaps more familiar in the religious than in the academic sphere.

The Argument in a Nutshell

Universally venerated by contemporary philosophers, the actual philosophic practice of Socrates is rejected or ignored. Socrates could never get a position today in a philosophy (or any other) department.

Socrates Tenured offers an account, and a critique, of the marginal role that academic philosophy plays in society. It focuses on an issue that has been ignored by the philosophic community: the institutional setting that philosophy has occupied since the creation of the modern research university. We see this setting – namely, the *department* – as the great unthought of contemporary philosophy.

Ours is both a theoretical and practical critique. Practically, this account constitutes a rejoinder to the ongoing reinvention of the modern research university. This reinvention, itself a response to various political and technological pressures, is likely to significantly affect both philosophy and the humanities more generally. Humanities departments at a few well-endowed schools will be able to continue on as before, but we suspect that for the vast majority of such departments, wrenching change lies ahead. The dangers we discuss, readily available for all to see, get little sustained academic attention for the simple reason that the people who set the agenda have a sinecure: tenure. This institutional fact has encouraged a blindness concerning the theoretical dimensions of our institutional housing.

Philosophers have a choice. We can continue to teach our classes and conduct our research while hoping that things continue on as before. Or we can try to take (partial) control of our future by changing our profession before our profession is changed by others. We choose the latter path, and call for the development of new institutional models that seek a way out from the cul de sac academic philosophy finds itself in.

1

But there is more at stake here than simply the fate of philosophy departments. The issue here is nothing less than the future of our fast-evolving society. Make no mistake: philosophizing always goes on in one way or another. Academic philosophy may have gone socially dormant, but new expressions of philosophy are blossoming across society. The dynamism of this modern-day Republic of Letters stands in stark contrast to the inward-looking conservatism of contemporary academics. This new Republic of Letters offers philosophizing on the fly, in response to a variety of game changers that have deeply philosophical elements – issues like climate change, artificial intelligence, globalization, new forms of media, and the potential remaking of the human genome.

These philosophers seldom hold PhDs in the field. They are usually identified, and self-identify, as business people, scientists, journalists, engineers, and futurists rather than philosophers. They have created an informal network of blogs, YouTube channels, magazines, and books. They function as freelancers, or inhabit a set of institutions that exist on the margins of the university, such as the Centre for Study of Existential Risk and the Future of Life Institute. Some of these institutions, such as X, formerly Google X, while on the margins of the academy, are very well placed in society. Other thinkers live a much more subterranean existence. In the aggregate, this network is fulfilling a role that academic philosophers have mostly abandoned – broad thinking in public venues about who we are, where we are going, and who we ought to be.

Socrates Tenured is a response to these developments. We first offer an account of the state of academic philosophy today, and of why contemporary philosophy (including applied philosophy) has failed to serve societal needs. We then provide an alternative model for philosophizing – what we call field philosophy – as a means to bridge the gap between (and temper the limitations of) academic philosophy and the amorphous, evolving techno-Republic of Letters.

Field philosophy has two basic features. First, it enlarges the range of academic philosophy by taking a more entrepreneurial approach towards philosophizing. We see field philosophy as a type of academic philosophy where one works on an ongoing basis with people in the STEM disciplines, the world of policy, community groups, and NGOs. On this account, field philosophy complements rather than rejects normal 'disciplinary' philosophy. In fact, we will argue for a circulatory model where field philosophers periodically return to the department to report on their experiences and to recharge their batteries before once again going out into in the world.

Second, field philosophy wants to bring the informal but intensely creative public forms of the modern-day Republic of Letters in closer relation with academic philosophizing. Philosophers within this techno-Republic of Letters would be enriched by a better acquaintance with academic work. They might, for instance, temper some of their optimism concerning the liberatory promise of accelerating technological advance. We believe that these two expressions of philosophy – the disciplinary work occurring in the academy, and the new Republic of Letters – need to recognize one another and function in a kind of informal partnership. Field philosophy can function as that bridge. And while we devote less attention to it, we will also note the existence of a third model for philosophizing – what we call the philosopher bureaucrat – in which academically trained philosophers permanently set up shop within extra-academic institutions across society. These three models – disciplinary philosophers, who mainly communicate with one another, field philosophers, who shuttle between academia and the larger world, and philosopher bureaucrats, who have 'gone native' – should constitute the ecosystem of twenty-first-century philosophy.

In sum, where others see gloom for philosophy, we see the chance for a renaissance – a rebirth of thoughtful action in an age that desperately needs it. And while some have claimed we are selling the soul of philosophy in order to comply with utilitarian accounts of value, we see ourselves as engaged in an act of jujitsu, turning the superior size of neoliberal forces to our common advantage.

Our argument can be boiled down to eight points:

1. Advocates of a market economy and technological progress have typically viewed their projects as the antithesis of philosophy, which they have dismissed as mere wool-gathering. Given the nature of academic philosophy, they have a point. But these social forces have now become philosophical in spite of themselves. Our global technoscientific culture raises any number of epistemic, ethical, hermeneutic, aesthetic, and metaphysical questions. This opens up new theoretical opportunities, as well as employment prospects, for philosophers both within and outside the academy.[1]

2. To seize these opportunities philosophers must interrogate their institutional setting. Treating philosophy (and the humanities generally) as a discipline – that is, as a regional ontology, consisting of specialists housed in departments – was the wrong response to the development of the modern research university. The exclusive disciplining of philosophy constitutes the original sin of twentieth and now twenty-first-century philosophy.

3. Addressing societal needs and responding to the challenges of neoliberalism requires new ways of philosophizing. What currently passes for philosophical research – as a sole model – is unsustainable: state legislatures will not continue to pay for research that is directed towards a small set of disciplinary peers. Our new cultural milieu also requires new models

for the teaching of philosophy. In both cases philosophy needs to be approached as a *practice*, with both research and teaching focused on spotting the philosophical moments residing within other disciplines, in social issues, in public and private institutions, and in everyday life.

4. A pluralistic approach to doing philosophy should also raise questions about what counts as quality philosophical thinking. Disciplinary philosophy has been uncritical about the question of rigour, part of its disregard of the philosophical dimensions of the field of rhetoric. Of course philosophers must make thoughtful and nuanced arguments. But what counts as excellence here should be treated as relative to the temporal, economic, and axiological needs of a given audience and situation.

5. The question of how to *implement* philosophical ideas needs to become part of our thinking – treated as a philosophical project in its own right. We need a research programme on the impacts of philosophy. This should in turn prompt the development of a general philosophy of impact that will be of interest across the academy (e.g. the STEM community) and to society at large. What does it mean for academic research to have an effect upon the world? What counts as a good effect? For at its root, the question of whether philosophy or any other research is useful or practical implies an understanding of what *counts* as useful or practical.

6. Twentieth-century attempts at philosophic relevance have had a melancholy fate. Applied philosophy, a creation of the 1980s that sought to make disciplinary philosophy more relevant, has largely been a failure – occasional exceptions notwithstanding. Rather than becoming a philosophical practice out and about within society, applied philosophy focused on writing philosophy articles for other philosophers. In contrast, field philosophy is attuned to the rhythms of contemporary society: practically engaged, stakeholder-centred, and timely.

7. Ours – we insist – is a plural agenda: traditional disciplinary philosophy is the source of valuable insight and should be supported. But it needs to be in dynamic balance with the more entrepreneurial approaches of the field philosopher and the philosopher–bureaucrat. The 110 PhD programmes in philosophy across North America should become experiments in different ways of practising philosophy and training philosophers – rather than the lemming-like repetition of the same that currently obtains in department after department.

8. A number of philosophers – a decided minority, to be sure – already practice something like field philosophy (and bureaucratic philosophy). Yet they rarely reflect on or write up their experiences, towards the goal of training the next generation of philosophers. This needs to change if their insights into different ways of philosophizing are to be built upon and institutionalized.

To describe matters synoptically, *Socrates Tenured* consists of three parts and encompasses four themes. In part I, chapter 1 offers a statement of the problem – the marginal societal role of academic philosophy, and a historical sketch of how we came to this juncture. In chapter 2, we describe the disciplinary status quo and outline the practical reasons why it is unlikely to last, by drawing together some of the numbers that illuminate the current situation of philosophy and the humanities more generally.

Part II (chapters 3 through 5) offers an analysis of various twentieth-century attempts to solve the problem of societal irrelevance. Chapter 3 looks at the field of applied philosophy writ large. Chapter 4 considers the case of environmental ethics. Both these areas are found to largely fail at the task of making philosophy relevant. We diagnose the lack of a larger societal impact in terms of 'disciplinary capture'. We then turn to bioethics in chapter 5. The situation here is different, and we provide an explanation for the field's relative success in having a broader impact.

Part III has two central points. Chapter 6 provides our resolution to the problem of societal irrelevance in terms of field philosophy. Chapter 7 is more prospective in nature: we consider the need to develop a new project across the whole of philosophy: the philosophy of impact.

To restate our four themes: we offer a description and diagnosis of the irrelevance of philosophy, an explanation of the generally failed attempts to be socially relevant, our own model for achieving societal relevance, and an outline of new areas for philosophic research.

To think with us, turn the page.

NOTE

1. To cite one example, the establishment of a 'broader impacts' criterion for the funding of grants by the NSF and other public science agencies around the world is a de facto acknowledgement of the central role of ethics and values within scientific research.

Prelude

Philosophy Purified

The story of Western philosophy can be told in a number of ways.[1] It can be presented in terms of periods – ancient, medieval, modern, and contemporary. Or as a quarrel between ancients and moderns, with 'postmodernity' somewhat awkwardly tacked on at the end. It can be rehearsed in terms of great thinkers: Descartes as the modern pivot, Frege (or Husserl) as having inaugurated twentieth-century thinking, Wittgenstein (or Heidegger) as the greatest thinker of the twentieth century. It can be seen as consisting of core areas (in the analytic tradition, metaphysics and epistemology and the philosophy of language) or traditions (in the continental tradition, phenomenology and existentialism). And the canon can be re-read in terms of gender and racial exclusion, as a discipline almost entirely fashioned by and for white men of European descent.

Yet despite the richness and variety of these accounts, all of them pass over a crucial juncture: the locating of philosophy within a modern institution (the research university) in the late nineteenth century. The institutionalizing of philosophy made it into a discipline that could be seriously pursued only in an academic setting. This fact represents the great unthought of contemporary philosophy.[2]

Take a simple detail: philosophy had never before had one central home. Philosophers could be found anywhere – serving as diplomats, living off sinecures, grinding lenses, even housed within a college or university. Afterward, if they were 'serious' thinkers, the expectation was that philosophers were inhabitants of the research university. Against the inclinations of Socrates, philosophers became experts like other disciplinary specialists. This occurred even as philosophers lectured their students on the virtues of Socratic wisdom, which highlights the role of the philosopher as the non-expert, the questioner, and the gadfly.

7

As Bruno Latour (1993) would have it, philosophy was purified. This purification occurred in response to two events. The first was the development of the natural sciences circa 1870 and the appearance of the social sciences in the decades thereafter. Each grouping consisted of successor disciplines to philosophy: in the first case out of natural philosophy, in the second case moral philosophy. The second event was the placing of philosophy as just one more discipline alongside these sciences within the modern research university. At the same time that learning was driven into the academy, philosophy lost its traditional position there, as the natural and social sciences divided the academic world between them.

This is not to claim that philosophy had reigned unchallenged within the university prior to the late nineteenth century. The place of philosophy had shifted across the centuries and in different countries. And many of those who we think of as philosophers did not work within the academy at all. But within or outside the university, 'philosophy' had always included the sense of being concerned with living a good life. Indeed, from the perspective of what followed (i.e. the development of a scientific research culture) earlier conflicts between philosophy, medicine, theology, and law were internecine battles rather than clashes across yawning cultural divides. These fields were thought to hang together in a grand unity of knowledge – a unity that shattered under the weight of increasing specialization by the end of the nineteenth century.

Early twentieth-century philosophers thus faced an existential dilemma: with the natural and social sciences claiming to map the whole of knowledge, what role was there for philosophy? There were a number of possibilities: philosophers could serve as

- synthesizers of academic knowledge;
- formalists providing the logical undergirding for research and education;
- translators integrating the disciplines and helping to bring the larger insights of the academy to the world at large;
- disciplinary specialists who focused on recondite philosophical problems in ethics, epistemology, aesthetics, and the like; or
- a combination of some or all of these roles.

But in terms of institutional realities, there seems to have been no real choice. Philosophers needed to become scientific. They needed to accept the structure of the modern research university, which consists of various specialties demarcated from one another. Real philosophers would be trained and credentialed as specialists. And so the 'discipline' became the reigning standard for what would count as proper philosophy. It was the only way to secure the field's survival.

But this description is misleading – because the kind of philosophy that predated this institutional shift did *not* survive. It's not like philosophy – that old creature – found a familiar niche in a new institutional ecosystem, one that would allow it to do what it had long been doing. Rather, philosophy itself changed, evolving into a disciplinary creature. The act of fitting (or fitting in) changed what survived. Philosophy was something new, even though most philosophers did not recognize the shift, preferring to believe that they and Socrates remained members of the same species. As if Socrates would stand a chance of surviving in the new institutional ecosystem.

This – the hiving off of philosophy as a discipline – was the act of purification that gave birth to the contemporary concept of philosophy. That is, to a degree hardly recognized, an institutional imperative drove the theoretical agenda. If philosophy was going to have a secure place in the academy it needed its own discrete domain, its own arcane language, its own standards of success, its own gatekeepers, and its own specialized concerns.

Having evolved into the same structural form as the sciences, it's no wonder that philosophy fell prey to physics envy and feelings of inadequacy. Philosophy adopted the scientific modus operandi of knowledge production – progress, rather than insight – but then failed to match the sciences in terms of advancement in describing (let alone controlling) the world. Much has been made of the inability of philosophy to equal the cognitive success of the sciences. But what has passed unnoticed is philosophy's all-too-successful mimicking of the institutional form of the sciences. We are judged by the same coin of the realm: peer-reviewed products. We too develop sub-specializations far from the comprehension of the person on the street. In such ways we are oh-so scientific.

Our claim, then, can be put simply: philosophy should never have been purified. Rather than being seen as a problem, 'dirty hands' should have been understood as the native condition of philosophic thought – a subject matter that is present everywhere, often interstitial, and essentially interdisciplinary and transdisciplinary in nature. Philosophy is a mangle. The philosopher's hands were never clean and were never meant to be.

There is another layer to this story. The act of purification accompanying the creation of the modern research university was not only about differentiating realms of knowledge. It was also about divorcing knowledge from virtue. Though it seems foreign to us now, prior to purification (and standardization, i.e., that philosophy is in principle no different from any other region of knowledge) the philosopher was assumed to be morally superior to other sorts of people. The eighteenth-century thinker Joseph Priestley wrote, "A Philosopher ought to be something greater and better than another man" (Priestley 1775, vol. 1, p. xxiii). Philosophy, understood as the love of wisdom, was seen as a vocation on par with the ministry. It required significant virtues (foremost among these, integrity and selflessness). What's more, in keeping with its questioning nature, philosophy was viewed more as a process than a product. The pursuit of wisdom was a virtue in itself, further inculcating these virtues. Knowing and being good were intimately linked, for the study of ethics elevated those who pursued it. The point of philosophy, after all, was to become good rather than only to collect or produce knowledge.

As Steven Shapin (2008) notes, the rise of disciplines in the late nineteenth century changed all this. The implicit democracy of the disciplines ushered in an age of "the moral equivalence of the scientist." The "de-magification of the world" (as Weber put it) put an end to any notion that there is something uplifting about knowledge. The purification of 'is' from 'ought' contributed to the feeling that it is no longer sensible to speak of nature, including human nature, in terms of purposes and functions. And by the late nineteenth century, Kierkegaard and Nietzsche had shown the failure of philosophy to establish any shared standard for choosing one way of life over another.

Once knowing and being good were divorced, the scientist and the philosopher could both be regarded as experts, but there are no morals or lessons to be drawn from their work. Science derives its authority from impersonal structures and methods, not the superior character of the scientist. The individual scientist is no different than the average Joe; he or she has "no special authority to pronounce on what ought to be done" (Shapin 2008, p. 25). Science became a "de-moralized" tool in the service of power, bureaucracy, and commerce.

Philosophy has aped the sciences by fostering a culture engaged in 'the genius contest'. Philosophic activity has become a competition to show how clever one can be in creating or destroying arguments. Like the sciences, philosophy has become a technical enterprise – the difference being that we manipulate words rather than genes. Lost was the once common sense notion that philosophers are seeking after the good life – that we ought (in spite of our failings) to strive to be model citizens and human beings. Having become specialists, we lost sight of the whole. The point of philosophy now is to be smart rather than good.

The irony today is that our culture's success at technical smarts has raised a whole new set of philosophical questions about the good. Which raises the question of whether academic philosophers are ready to help society think.

NOTES

1. This prelude is a revised version of: Frodeman, Robert, and Adam Briggle (2016), 'When Philosophy Lost Its Way', *The New York Times*, January 11.

2. We discovered, just as this was going to press, that Loncar (2016) raises a similar question: "What is the philosophical significance of academic disciplines and philosophy's inclusion within disciplinary structures?"

Part I

PHILOSOPHIZING IN NEOLIBERAL TIMES

Chapter 1

Philosophy, Know Thyself

There are nowadays professors of philosophy, but not philosophers.

Thoreau

One of our professors once explained the nature of the dissertation and one's subsequent philosophical career:

> Find a small topic that no one has studied; dig into it for a few years. Don't stop until you know it better than anyone – except for the other 30 or 40 specialists in the area. When you graduate you will then spend the next 30 years writing for that group of specialists. That's how it works.

It's time that we question our allegiance to this model of philosophy.

THE PHILOSOPHER AND THE POLIS

Everyone struggles to live a rich and fulfilling existence – even, or especially, as many of us are surrounded by an abundance of means and opportunities. But frame the question in terms of philosophy and the conversation goes stale. People grow impatient: philosophy is fine in the seminar room, but in the real world such abstractions are dismissed as wool-gathering. It's a contradiction: people describe ethics as merely a matter of opinion even as they struggle to ensure that their children are treated fairly. And they dismiss aesthetics as subjective even as they plan trips to national parks and pour over the details of their kitchen remodel. Philosophy is impractical – and unavoidable.

To philosophize today – by which we mean professionally, in a salaried position, at a college or university – is to live within this paradox. One could claim that it has always been so, that a gap has always existed between the concerns of philosophers and our real-world philosophic problems. Tension between the language of philosophers and the philosophical dimensions of everyday life has been part of our cultural DNA since the milkmaid laughed at Thales tumbling into a ditch.

Of course, Thales had practical chops too, which he showed by making a killing in the olive market. Nonetheless, the vexed relationship between philosophers and society is a perennial fact. The most famous example, of course, is given by the fate of Socrates. Philosophy has always alternated between boring and irritating the outside world. It is a tension that philosophers have sometimes cut, Gordian knot-like, by retreating into abstruse speculation. But this has always prompted a countermovement decrying the irrelevance of philosophy. Chief among those complaining about the uselessness of philosophy have been philosophers themselves. Thus, Descartes scorned the abstractions of the Schoolmen and Marx said the point of philosophy was to change rather than merely interpret the world.

If the relationship between philosophy and the polis has always been fraught, and perhaps laced with a bit of subterfuge, it has also been in the end a workable one. Until the twentieth century, that is. Since then the tension has grown into a paradox, the gap into a chasm. *Socrates Tenured* offers an account of the development of this chasm – how philosophy, the most practical (if not always the most efficient) of subjects, lost the creative tension between contemplation and engagement and slipped into cultural irrelevance. We also offer more than critique: we propose a way forward, describing how philosophy, especially philosophical research, can regain a role in culture.

Our argument focuses on the single greatest impediment to philosophy's societal relevance: the emergence of the field as a *discipline*. The early twentieth-century research university disciplined philosophers – or more precisely, given their limited set of options, philosophers chose to discipline themselves. Philosophers were placed in departments. They inhabited libraries and classrooms. Their writings were restricted to professional diction and concerns. And they wrote for and were judged by their disciplinary peers.

William James was among the few to notice this shift in the circumstances of philosophy. As early as 1905 he lamented the "desiccating and pedantifying process" that philosophy had become. He had a word for it: "Faugh!" The young philosophers then coming of age regurgitated "what dusty-minded professors have written about what other previous professors have thought" (Bordogna 2008). Such was the birth of the academic discipline of philosophy, which, like a snake, turned and swallowed its own tail.

In the main, however, these changes were little noticed. They were treated as merely the matter-of-fact stuff of being 'rigorous' and 'serious'– the professionalization of the field. The fact that this constituted a new material culture that institutionally speaking turned philosophy into a regional ontology was passed over in silence. It continues to be passed over today. Like Moliere's Gentleman, to whom no one had explained that he had been speaking prose, philosophers seemed innocent of the fact that they had been disciplined. Or that they might have reasons to object to this fact. Their field of play now consisted of texts; à la Kuhn, their work consisted of puzzle solving or 'normal philosophy'. Everything outside their research portfolio counted as the 'gossip' of conference dinners after the real work of listening to one another's papers had been completed. Matters like being housed in individual offices next to other philosophers, the job possibilities of their graduates, or the philosophic dimensions of the day-to-day work of private firms and public agencies, while perhaps interesting, were peripheral to the real stuff of philosophy. The institutional trappings of the field were treated as simply the banalities necessary to provide a space for the pure flower of philosophy to bloom.

In hidden ways, however, these 'banalities' have structured and directed the theoretical content of philosophical work. They've helped define standards of rigour, suitable topics and styles of discussion, and appropriate audiences. The purity and neutrality of the philosophical enterprise, in other words, has been neither. The field has assimilated its social and material conditions in a dogmatic manner. Philosophy has become the creature of twentieth-century disciplinary culture.

We seek to revive a Socratic practice. The first step towards doing so is to note the obvious: that Socrates was not an employee of an institution of higher education, but was privately employed, supporting himself as a stone mason, while spending many of his days hanging out in town talking to all kinds of people. The second step is to note that if he went by any other name, Socrates wouldn't be hired today by any department of philosophy: his way of practising philosophy would be dismissed as hopelessly amateur. Socrates is venerated, but not taken seriously.

We think this is a problem.

The response to Socrates hasn't always been veneration. In our period Socrates has been presented as a slightly cranky saint. But in earlier times he was taken seriously enough to be worth criticizing. Denunciations began during his lifetime (e.g. *The Clouds*); he was charged with impiety and corrupting the youth, accusations that, even setting aside the political context, were plausible enough. In the years after his death, Socrates has had a wealth of additional critics: Aristoxenus wrote a *Life* said to be even more censorious than that of the accusers at his trial (Morrison 2011). And the grievances of the Epicureans were substantial, questioning Socrates' claim that virtue cannot be taught and describing him as a sceptic and sophist.

The Platonic texts were lost with the fall of Rome, and so Socrates largely faded from view. With the Renaissance and the rediscovery of Plato, the hagiography grew: Mill was one of many who compared him to Jesus, that other victim of a judicial murder. Nietzsche presents the great counter-instance. He criticized Socrates as a decadent who rejected the truth of the senses to devote his life to abstract theorizing: "The most blinding daylight; rationality at any price; life, bright, cold, cautious, conscious, without instinct, in opposition to the instincts" (*Twilight of the Idols*, section 11). But we aim at neither hagiography nor condemnation. We simply want Socratic practice to be taken seriously as one possible model of the philosophic life.

Today the institutional trappings of disciplinarity have built a wall between philosophy and its social context. Even when their subject matter consists of something of real significance to the wider world, philosophers discuss topics in a way that precludes the active interest of and involvement by non-philosophers. Philosophers may have much to say to their fellow citizens, but unlike Socrates they no longer tarry in the *agora* to say it. It's pretty ironic for a profession premised on the dictum 'know thyself'. One finds no explorations of the effects that the department might have had on philosophical theorizing, or of where else philosophers could be housed, or of how philosophers, by being located elsewhere, might have developed alternative accounts of the world or have come up with new ways and standards of philosophizing. When philosophers leave behind their disciplinary habitats, living and working elsewhere than in philosophy departments, the standards for their work change. For when you change the place, and the audience, you change the criteria for determining what counts as 'real' or 'good' work.

Philosophers once recognized that their work is not simply one discipline alongside the others. It was understood that in addition to fine-grained analyses, philosophy offered perspectives that undergirded, capped off, or synthesized the work of other disciplines such as physics or biology, and then connected those insights to our larger concerns. Such work lost favour in the twentieth century – dismissed as Weltanschauung philosophy by analytic philosophers, and as foundationalism by continental philosophers. But reopen this perspective and questions abound: If philosophy is not, or not exclusively, a regional ontology, why are philosophers housed within one region of the university? Why is peer-reviewed scholarship the sole standard for judging philosophic work, rather than also considering the impacts that such work has on the larger world? Why, for instance, when we want to rank philosophy departments, do we only ask other philosophers – and ones at the so-called top universities at that? And why are there only two social roles for those with PhDs in philosophy – to teach undergraduates, and to talk to other PhDs in philosophy?

Michael Rinella notes that, in an interview with Paul Rabinow shortly before his death, Michael Foucault spoke of his concern with "what one could call the problems or, more exactly, problematizations" – how we decide what questions do or do not get asked. As Rinella (2011) puts it:

> Over time a domain of action previously accepted as given evolved into something deemed worthy of sustained critical commentary, often in association with particular social, economic, or political processes.

By what trick of intellectual history have philosophers not asked questions about their disciplinary status? Or developed robust accounts of the societal impact of their research, rather than relying on hackneyed accounts of the virtues of critical thinking? And so we seek to problematize a series of questions that philosophers have passed by without notice.

THE PROBLEM

Compare two quotes, from two of the most prominent thinkers of the twentieth century:

> Think of organic chemistry; I recognize its importance, but I am not curious about it, nor do I see why the layman should care about much of what concerns me in philosophy.
>
> Quine

> Philosophy recovers itself when it ceases to be a device for dealing with the problems of philosophers and becomes a method, cultivated by philosophers, for dealing with the problems of men.
>
> Dewey

These quotes represent not just two different attitudes, but two different models for how (and where and with whom) to conduct philosophical thinking.

In 1917, John Dewey published 'The Need for a Recovery of Philosophy', a reflection on the role of philosophy in early twentieth-century American life. In it, Dewey argued that philosophy had become "sidetracked from the main currents of contemporary life," too much the domain of professionals and adepts. He took pains to note that the classic questions of philosophy had made immense contributions to culture, both past and present. But he was concerned that the topics being raised by the new class of professional philosophers were too often "discussed mainly because they have been discussed

rather than because contemporary conditions of life suggest them." Dewey
soon travelled to China, where he delivered nearly 200 lectures on education
and democracy to large crowds across a two-year stay. Upon his return, he
continued to comment on the public questions of the day, a role he filled until
his death in 1952. But since then another set of expectations has come to rule
the philosophic community.

The reasons for this shift are open to debate. Jacoby (2000) chalks it up
to the allure of academic careerism during the post-World War II expansion
of universities. McCumber (2001) notes the chilling effects of McCarthyism.
Reisch (2005) sees it as largely a matter of historical accident – who survived
the war years to set the direction for post-war philosophy. Soames (2005)
describes it as the playing out of the logic of specialization. And we will
argue that it is largely a matter of the consequences that flow from the uncritical embrace of a certain institutional housing.

But whatever the cause, over the course of the twentieth century, philosophers increasingly abandoned the way of Dewey to follow Quine's path in
treating philosophy as a technical exercise of no particular interest to the
public. While it is possible to point to philosophers who *work with* (rather
than merely talk about) non-academics, among the mass of philosophers a
lack of societal engagement is treated as a sign of intellectual seriousness.[1]
As Quine put it in a 1979 *Newsday* piece, the student who "majors in philosophy primarily for spiritual consolation is misguided and is probably not
a very good student." For Quine, philosophy does not offer wisdom; nor do
philosophers "have any peculiar fitness for helping ... society." It's hard to
imagine a less Socratic approach to philosophy. Even more surprising is the
historical amnesia where such an attitude goes unchallenged by the philosophic community.

We live in a global commons both created and constantly modified by
technoscientific invention. We are surrounded by events crying out for philosophic reflection. The issues are incredibly varied: the creation of autonomous killing machines, the loss of privacy in a digital age, the introduction of
driverless cars, a changing climate, the remaking of friendship via Facebook,
the refashioning of human nature via biotechnology. What unites these cases
are the philosophic (at turns, ethical, political, epistemic, aesthetic, and metaphysical) questions they assume or provoke.

It is certainly possible to find thoughtful explorations of such questions.
They appear in any number of books, and in magazines and blogs – in places
like the *New Yorker*, *Wired*, and *Three Quarks* – constituting a latter-day
Republic of Letters. In this sense, as Romano (2012) and Goldstein (2014)
have argued, philosophy abounds. But the situation remains skewed: when
we turn to academics for help with societal problems it is the scientist and the
economist that we call, not the philosopher. In the United States today there

are some 15,000–20,000 PhDs trained in one or another aspect of philosophy; but they overwhelmingly communicate only to one another (and with students, of course, in the classroom setting). In fact, it is taken as a sign of their professionalism that they do so.

Karl Jaspers (1963) noted a half-century ago (in *The Atom Bomb and the Future of Man*) that science, technology, and economics have overcome themselves: technical issues have morphed into a series of philosophic questions. Our growth-oriented, materialist lifestyle has lifted us out of poverty, but does not satisfy our most fundamental needs. In fact, further economic and technoscientific progress now threatens to trivialize (and possibly destroy) us. Academic philosophers have a distinctive set of skills and perspectives to bring to these questions. But they have been missing in action.

The problems with academic philosophy are both theoretical and political. Consider first the theoretical side of things. Take the field of metaphysics; every philosophy department teaches the course. But how is the subject handled? Google 'metaphysics syllabus': the situation remains as Dewey described it. Classes begin from the concerns of philosophers rather than from contemporary problems. Consider as magisterial a source as the *Oxford Handbook of Metaphysics* (2003) edited by Loux and Zimmerman. Their introduction begins:

> Its detractors often characterize analytical philosophy as anti-metaphysical. After all, we are told, it was born at the hands of Moore and Russell, who were reacting against the metaphysical systems of idealists like Bosanquet and Bradley.

It's a discussion framed in terms of a roll call of philosophers. One finds no reference to people's actual lives, to the metaphysical issues tied to the births and deaths, inventions and transformations that surround us. There is no acknowledgement that metaphysics consists of some of the most intimate and portentous questions of our lives. Instead, it is a tale told in terms of professional thinkers and their problems: Moore and Russell, Bosanquet and Bradley, Quine and Lewis.

The eight sections of the *Oxford Handbook* continue this Olympian perspective:

- Universals and Particulars
- Existence and Identity
- Modality and Possible Worlds
- Time, Space–Time, and Persistence
- Events, Causation, and Physics

- Persons and the Nature of Mind
- Freedom of the Will
- Anti-Realism and Vagueness

Chapter titles erect barriers to the uninitiated: 'Supervenience, Emergence, Realization, Reduction' and 'Compatiblism and Incompatiblism'. Now note, we are *not* claiming that the matters addressed by such essays are insignificant. But it takes an expert in philosophy to extract the nut of existential meaning from the disciplinary shell. Evidently, it is not considered part of the philosopher's remit today to ground questions of universals in the everyday lifeworld. And so philosophers discuss the issue in terms of 'redness' rather than 'Caucasian' or 'gender'. Whatever happened to Hegel's sense that philosophy needs to extend a ladder to where people are? Or that the abstractions of philosophy are motivated by the most existential of issues? It is no wonder that so many walk away confirmed in their prejudices concerning the irrelevance of philosophy to everyday life.

Philosophers begin with insider topics, but metaphysical issues are in the news every day. A recent newspaper article describes a patient taking heart pills that include ingestible microchips: the chips link up with her computer so that she and her doctor can see whether she had taken her meds. The story also describes soon-to-be marketed nanosensors that will live in the bloodstream and be able to spot the signs of a heart attack before it occurs. These are issues that fall under 'Existence and Identity', one of the sections of the *Oxford Handbook*. At stake here is not just new physical instrumentation, but also questions about the nature of self and the boundary between organism and machine. Loux and Zimmerman miss their chance to frame this section in terms of our increasingly Borg-like existence rather than solely in terms of scholastic debates.

But to ask a philosopher *why* these categories are important, or how they tie into our lifeworld, is to brand yourself as not serious about philosophy. Such high-toned dismissal would be laughable if it were not both dogmatic and self-destructive. Nor is this only a problem of contemporary analytic philosophy. You might think that continental philosophers, heirs to the tradition of existentialism and phenomenology, would present a different face to the world. But they practice their own form of disciplinary pathology, often expressed in terms of derivative accounts of leading thinkers. Heidegger criticizes philosophy for the forgetfulness of being, but neither he nor his followers apply the point to life within the modern research university, – that philosophy itself has become, institutionally speaking, a regional ontology, having forgotten to ask questions about its own particular mode of existence as a self-referential exercise among philosophers.

THE NEOLIBERAL ELEMENT

Now consider the political side of things. In January 2015, just prior to announcing his candidacy for president, Wisconsin governor Scott Walker stated that university professors should be "teaching more classes and doing more work." Walker also sought to rewrite the University of Wisconsin mission statement, inserting "meet the state's workforce needs" while cutting "search for truth" and "improve the human condition" (the attempt was stymied after news reports). By June of that year Walker had cut the University of Wisconsin system budget by $250 million. At the same time he signed into law provisions that removed tenure protections from state statutes.

Similar moves are afoot across the nation and, indeed, in the UK and other parts of Europe as well. The research university is becoming politicized in ways foreign to earlier generations. Professors are increasingly viewed as simply another (liberal) interest group that, like labour unions, feather their beds at public expense. The days are gone when the pursuit of truth served as sufficient justification for supporting academic research. Like climate science, the university is seen today as a bastion of democratic politics, fair game for Republican legislatures seeking to shrink the public sphere. Across the bulk of the twentieth century, conservatives had as much reason to support the "search for truth" as liberals. Capitalism had faltered in the Great Depression, and across World War II and the Cold War great ideologies were in contention. Today, however, a generation past the fall of the Berlin Wall, this philosophical debate seems settled; capitalism stands triumphant. Politicians vie with one another in bringing market mechanisms and attitudes to campus while stripping away the privileges and protections professors once enjoyed.

Additional examples abound. In 2015, a North Carolina legislator introduced a bill requiring professors in all fields at every state university (including the flagship, the University of North Carolina, Chapel Hill) to teach a 4-4 schedule. In other words: no more research. The bill never made it out of committee; but the glib responses we hear from our colleagues, that laws like this will simply make it easier for other universities to snatch that state's best professors, badly misread the situation. The wholesale elimination of research is unlikely to occur; the STEM disciplines (science, technology, engineering, and math) can make arguments concerning their contributions to health care, technological advance, economic growth and the like. But this isn't the case with philosophy and the humanities.

And so we should expect legislators to latch onto reports that note the abysmal rates of citation for research in the humanities. In one study, of the thirteen research articles published by SUNY-Buffalo literature professors in

2004, eleven received between zero and two citations, one had five citations, and one had twelve (Bauerlein 2011). Another analysis found that 82% of peer-reviewed publications in the humanities are never cited. The study's authors estimate that less than 5% of peer-reviewed humanities papers are ever read by anyone (Biswas and Kirchherr 2015). Politically, it matters little if these numbers are correct; while scholars debate their merits, the numbers will be used in partisan wars seeking to defund the university, eliminate tenure, and cut research time for humanities professors.

What can be said in defence of the scholarly status quo? Publications, whether cited or not, bring the pleasure of intellectual exploration to their authors. They also provide a means (specious or not) for distinguishing between candidates for tenure and promotion. But increasingly the question will be asked – and not only by conservatives – why should taxpayers and students subsidize reams of unread scholasticism? Or as Bauerlein puts it, "After 5,000 studies of Melville since 1960, what can the 5,001st say that will have anything but a microscopic audience of interested readers?"

Philosophers and humanists have different responses to these criticisms. Geoffrey Harpham, then director of the National Humanities Center (2011), for example, defends the 5,001st article on Melville – or as he frames it, the 16,772nd book on Shakespeare – by arguing that as times change we bring new eyes to old texts, finding new insights. Surely this is correct. But he does not explain how these new insights are valuable if no one is reading them. There are other replies, of course: research hones teaching skills; citation counts are crude metrics of value; humanities research contributes to the improvement of culture; and the occasional breakout publication causes a stir.

These aren't terrible responses. But they are incomplete, and are presented in too pat a manner. The issue is often framed as a matter of justifying "the value of the humanities," especially research in the humanities (e.g. Small 2013). But it's too easy to talk about 'values' without any actual reference to specific situations. Just trot out some well-worn phrases about sweetness and light. Given the era we live in, philosophy needs to be treated as a practice. More than simply talk about values, we need an account of how those values are implemented and mobilized. This itself is a philosophical task (see chapters 6 and 7) that calls into question the disciplinary model of knowledge production, where a philosopher's primary audience consists of other philosophers, and requires careful reflection on the use of metrics, indicators, and narratives. Rather than 'value', we should be thinking about impact.

This will be a better way to deal with the 'show me the money' mindset storming the academy. The account we offer here, focused on philosophy but hopefully relevant across the humanities, seeks to respond to the main factors lying behind these changed circumstances: the democratization of knowledge, the growing force of neoliberalism, the development of an audit culture, and the rise of academic metrics – at a social moment when philosophic perspectives have never been more required.

We expect that changes to philosophy departments will move from less established institutions (e.g., 'directional' universities such as our own) upward to more prestigious universities (cf. Christensen's 'disruptive innovations'). Pat answers may well be sufficient at elite universities, at least in the short term, for such schools often have the resources that allow them to ignore the philistines in state legislatures. But if philosophical research is to broadly prosper we will need far more than the tired defences of the last 100 years. And so we offer a de-disciplined and post-disciplinary model of philosophy, a philosophy that is ready to be taken into the field. For unless professional philosophy embraces *and institutionalizes* an engaged approach to philosophizing, working alongside other disciplines and abroad in the world at large, its future is likely to be one of ever further marginalization. One hundred years after Dewey's essay, it is time for another recovery of philosophy. But not simply a recovery; we need innovations and a more entrepreneurial expression of philosophy. For the academy, the last of the handcraft industries, is about to be transformed.

MODE 2 PHILOSOPHY

As one of us has argued elsewhere (Frodeman 2013), the broader societal meaning of 'interdisciplinarity' is that disciplinarians of all stripes now have to justify their existence to a wider set of peers. Thus, in *The New Production of Knowledge* (1994) Gibbons et al. chronicled the shift in late twentieth-century science from "Mode 1" to "Mode 2" knowledge production. Mode 1 is classically academic, investigator-initiated, and discipline-based research. Mode 2 is context-driven, problem-focused, and interdisciplinary (better, transdisciplinary) in nature. To adopt that language for the moment, we are promoting the development of Mode 2 philosophy.

Make no mistake: we are pluralists on this point. Mode 1 or disciplinary scholarship should continue to play a central role in philosophy. As Kitcher (2011) notes, there are many places in contemporary philosophy where important work is being done:

Philosophers of the special sciences, not only physics and biology but also psychology, economics, and linguistics, are attending to controversies that bear on the future evolution of the focal field, and sometimes on matters that affect the broader public. Some political philosophers are probing the conditions of modern democracy, considering in particular the issues that arise within multi-cultural societies. Ventures in normative ethics sometimes take up the particular challenges posed by new technologies, or the problems of global poverty. Social epistemology has taken some first, tentative, steps. A growing number of thinkers are engaging with questions of race, gender, and class.

This constitutes work of interest in its own right, and which over time disseminates into other disciplines and into the culture at large – what we will call the trickle-down model. But such efforts need to be complemented by an equal focus on work that is socially engaged. Not commentary *about* societal problems, but rather actual presence in the field, engaged in an ongoing, day-to-day or week-to-week fashion with non-philosophers. In part, this is simply a matter of recognizing a new reality: society is demanding that academics demonstrate their broader relevance. Philosophy needs to demonstrate its bona fides by showing how it can make timely and effective contributions to contemporary discussions. But this is also in recognition of a basic fact: philosophy needs to get outside more often. The sunshine will do it good.

This shift won't be easy. It will require serious philosophical thought – about how to practice philosophy on the fly, and how to hone one's rhetorical chops. Towards that end, it's instructive to compare notes with policy studies, a field that has been a blind spot for philosophy. In our experience, philosophers are often not even aware of the field's existence. And when so apprised they often dismiss its concerns as having already been addressed by social and political philosophy, or as merely consisting of alternatively 'bureaucracy' or 'activism'. This is a serious error. Policy studies has carved out a crucial niche: housed in the academy, but positioned between the disciplines and the wider world. The field is concerned with how decisions are made and how knowledge is taken up (or not) in the making of policy. The success of philosophy at becoming more socially engaged will in part turn on integrating the insights of policy studies into its worldview (Frodeman and Mitcham 2004).

Policy studies asks how scientific research is translated into social improvements or 'impacts'. Throughout the Cold War the answer to this question was basically serendipity. In *Science, the Endless Frontier*, a 1945 pamphlet presented to President Truman and written at the request of Franklin Roosevelt, Vannevar Bush (who led scientific R&D efforts during the war) argued that basic research is foundational to social progress. We know science will improve society, but we don't know which research will lead to which improvements. There is no way to predict how things will turn out. So it is best to conduct as much basic research as possible, as widely as possible, trusting that somewhere down the line it will pay off. This faith in serendipity allowed scientists to wall off a narrow domain of responsibility: their job was to conduct quality research as judged by their disciplinary peers. The results were then "thrown over the wall" to society – a phrase that one of us heard repeatedly when working with the US Geological Survey. The uses that such research was put to were not their responsibility.

Philosophy also took on these institutional assumptions. What do we have in philosophy departments if not, in Bush's words, "the free play of free intellects, working on subjects of their own choice, in the manner dictated by their curiosity"? There is the same guiding assumption that somehow, somewhere down the line all those words in peer-reviewed philosophy journals will pay rewards to society. Philosophy plays the long game, razing conceptual structures and replacing them with new ones. And this is not the sort of project that can be rushed. The occasional philosopher has articulated how this kind of 'trickle-down' model of philosophy benefits society. But it has remained a tacit article of faith.

That faith is no longer sufficient. At the federal level, budget cuts and a growing animosity towards the public sphere have led to Congressional attacks on individual research grants and even entire research programmes (the fate of the Social Science Directorate at the NSF hangs in the balance). This has also led to attempts to measure the broader impacts of research through the development of bibliometrics, economic analyses, and Altmetrics (a topic we have been working on). But while such questions are a hot area of research within policy studies, philosophers and other humanists still plod along with the serendipity model embedded within Mode 1 disciplinary research. They have not yet taken note of these changed conditions – or if noted, seen them as an opportunity for interesting theoretical work that also holds the possibility of important practical outcomes.

It's more than time for a 'philosophy policy' or a humanities policy analogous to science policy. Indeed, we believe that this is already beginning to germinate. While Mode 1 philosophy is still the reigning orthodoxy, there is a growing countermovement within the ranks of philosophers, sometimes lumped under the title of 'public philosophy'. We call our own version of Mode 2 work 'field philosophy'. There are a number of similar approaches in areas such as environmental justice, critical race theory, feminism, and bioethics work that we recognize as allies (chapter 6). We celebrate these diverse approaches to Mode 2 philosophizing. But we believe that the lack of thought given to the institutional dimensions of philosophizing has limited the effectiveness of this work. A new philosophical practice, where philosophers work in real time with a variety of audiences and stakeholders, will lead to new theoretical *and* institutional forms of philosophy – once we break the stranglehold that disciplinary norms have upon the profession.

OUR PEER COMMUNITY

Given these goals, we have a number of audiences in mind for this book. For administrators, scientists, engineers, and others outside the discipline, our goal is to introduce a kind of philosophy that shatters preconceived notions of philosophers as navel-gazing nook-dwellers. For philosophers our goal is to open up new opportunities for theorizing, social engagement, and employment. Quite often, when philosophers follow the urge to engage in real-world problems, they wind up working through a set of thorny theoretical and practical issues with no resources to help them. We hope this book serves, if not as a set of best practices, at least as a reference that contextualizes these challenges and as a vehicle for promoting the creation of an enduring community of practitioners. In the absence of such a self-reflexive community, experiments in Mode 2 philosophizing will remain a series of one-offs. Isolated individuals fed up with the disciplinary status quo will reinvent the wheel of alternative philosophical practice. Lessons will be learned but not shared. Traps that could be avoided will be fallen into, and alternative career paths for philosophers will remain at the margins.

For make no mistake: it will take a community to institutionalize Mode 2 philosophic practices. As it stands now, heterodox practitioners (however they may self-identify) lead professional lives that run against the grain. As Linda Martín Alcoff notes, many Mode 2 philosophers try to "walk a fine line between responsiveness to community needs and employment survival, pushing the boundaries of academic respectability even while trying to establish their credentials in conventional ways" (Alcoff 2002, p. 522). It is these "conventional ways" that must change. We need a philosophy where responsiveness to community needs, rather than only disciplinary interests and imperatives, is an integral part of one's employment and is viewed as academically respectable.

This is how we can shrink the chasm between philosophy and society. This will require a number of institutional changes, from revised curricula to modified promotion and tenure criteria to alternative metrics for excellence and impact. As these changes are implemented, it will be important to consider at what point the chasm has been reduced to a right-sized gap. After all, we don't want to eliminate the space between philosophy and society altogether. Philosophy remains a contemplative exercise. Socrates was engaged, but still an outsider. He certainly was no pundit looking to score the most outrageous sound bite and rack up the most 'likes' on Facebook. We need a people's philosophy that reserves the right to be unpopular.

But if this is in part a book for philosophers, we admit to having some concerns about this particular audience. It is far from clear that philosophers feel the need to be saved from the error of their ways. We find ourselves in the uncomfortable position of offering a solution to a crisis that most of our colleagues do not think exists. Philosophers who are safely ensconced in a tenured or tenure-track position have long been rewarded for their disciplinary bona fides. They are secure, and are in no mood for changes that strike them as unserious and contrary to the disciplinary virtues they have grown to appreciate.

Then again, it may be that philosophers are starting to awaken to the challenges we find so pressing. A 2015 survey[2] conducted on the Leiter Reports (a prominent philosophy blog) asked, "What are the ten most pressing issues" facing the profession? In total, 725 philosophers responded. Their top four concerns are telling: (1) "Bad job market, decline of tenure-track faculty positions, increasing reliance on adjuncts," (2) "Declining state support for higher education," (3) "Hyper-specialization and/or increasing irrelevance of philosophy to public/culture at large," and (4) "Erosion of tenure." As one commentator noted about the discussion sparked by the survey (Arvan 2015), there is very little written about what can or should be done to address these top concerns (strikingly, despite its ranking (tenth), most discussion focused on questions of gender equity, an area that has garnered public attention). We offer this book as a catalyst for thinking about and doing a new kind of philosophy in an age of dismal academic job prospects, shrinking budgets, and increased accountability.

Still, we recognize that our argument may fall into the no man's land between the vast majority of philosophers who will find it outlandish, and a minority who will claim that this is old news. For example, one of our suggestions is that philosophers need to be housed in biology, chemistry, and engineering departments. Most philosophers will find this to lie somewhere between the unpalatable and the insane; but there are a few who will point out that they already happily live and work within other departments. Our reply to the minority of those who are already practising Mode 2 philosophy is that institutionalizing such alternative practices is the difference that can make all the difference. Mode 2 philosophy needs to mature from a band of irregulars to a legitimate institutional presence. Our reply to the majority is that they are failing to see how precarious their position actually is. The winds of change are blowing (just ask the graduate students looking at a grim academic job market). How much longer can the disciplinary orthodoxy survive at institutions that lack a billion dollar endowment?

We do want to challenge one misconception. The argument made here is sometimes viewed as mercenary, utilitarian, or vulgar in nature, a sell-out to neoliberalism, or the denigration of all that is fine and beautiful in philosophy. We view this as wrong on two counts. First, it misses that there is something fine and beautiful about mixing it up in the public square. At least Socrates thought so; so did Leibniz, Marx, Russell, and Dewey. Philosophy is not such a delicate flower that it cannot take a little pummelling. Second, the public square itself is in disrepair, and in desperate need of more contemplative moments. One of the casualties of an inwardly focused, disciplinary philosophy has been the paucity of accounts that defend the contemplative element of life, *in situ*, *in media res*. Bringing philosophy into the public sphere offers a chance to highlight utilities other than economic ones. The engaged philosopher need not sell out to utility. He or she can also redefine it.

NOTES

1. By *working with* we mean something other than – and much more hands-on than – that legendary beast the 'public intellectual'. We will also distinguish our approach from applied philosophy, which largely consists of *writing about* problems rather than working with people to solve them.

2. http://leiterreports.typepad.com/blog/2015/03/readers-identify-the-most-important-issues-in-the-profession.html.

Chapter 2

The State of Things

Rather than being an innovation, our approach has an ancient if neglected pedigree. Socrates practised a situated philosophy, moving throughout society and striking a blow for thoughtfulness whenever the opportunity arose. This was an unstable and dangerous practice, but one that remained hitched to the lifeworld. Academic philosophy today is far more comfortable. But it has also rendered itself largely irrelevant by ignoring the priors of its own situation.

Plato's *Republic* offers *avant la lettre* criticism of disciplinary philosophy. At the start of Book 2 Socrates labels the preceding conversation a preamble (*pro-omion*, the way or path before), the work that's necessary before turning to the main argument. Book 1 addresses the challenge that precedes every act of philosophizing: in a given situation, how much philosophizing is possible? For one must have the *opportunity* to philosophize. In Book 1 Socrates either tames or runs off Cephalus, Polemarchus, and Thasymachus, the representatives of a life lived on the basis of piety, tradition, and self-interest. But the larger point here is that Socrates had to acknowledge and address these positions, rather than simply begin thinking. In the real world, philosophers always find themselves constrained by one or more of these factors.

The philosopher's task, then, is not simply to come up with insights, but to also determine how much philosophizing a given situation can stand before thinking becomes impossible or counterproductive. But this is precisely the point that the 'department' allows philosophers to ignore. The only constraints to philosophy within the department are ordinary ones – the length of the seminar or the semester, word counts, or page limits. Whereas outside the department, even if we set aside intimidating situations, debate among even well-intentioned adults seldom involves more than a few back-and-forths before bumping into personal or political biases, non sequiturs, *ad hominum* attacks, and the pressures of time or money.

The department is a space mostly outside of such concerns – a luxury which we heartily appreciate in our own lives. But it should not blind us to the fact that, like the walled-off scientific experiment, there is an important degree of irreality to the department.

THE CRISIS MOTIF

As we have noted, the argument of *Socrates Tenured* operates on two levels, the theoretical and the practical. In terms of the former, we are arguing that academic philosophy has become untrue to its own best nature by becoming an unnecessarily technical exercise mostly of interest to other professional philosophers – even as the world abounds with problems that have philosophical dimensions. Our way forward, which is the focus of part III, is to institutionalize an interdisciplinary and transdisciplinary way of doing philosophy, what we call field philosophy.

But our argument isn't built solely on a theoretical account of the deficiencies of contemporary philosophy. In what follows, we provide an additional description of exogenous forces that are likely to compel philosophers to adopt a new, transdisciplinary way of doing things. This chapter draws together data on the state of the profession in support of our account. Of course, data rarely compels a decision or a change of heart. But it does give the reader a second set of reasons for taking our approach seriously.

The crisis motif surrounding philosophy, and the humanities generally, has become a familiar genre. Articles appear with regularity in the popular media (*The New York Times*, *The Atlantic*, *The Chronicle of Higher Education 'Review'*), the blogosphere (*3 Quarks*, *Scientia Salon*), as well as in scholarly articles and monographs (Kitcher's 'Philosophy Inside Out'; Nussbaum's *Not for Profit*, Donoghue's *The Last Professors*). We've made our own contributions to the genre (Briggle and Frodeman 2011; Briggle, Frodeman, and Barr 2015).

Diagnoses take different forms, and operate at different levels. One type is familiar to students of philosophy: the philosopher's announcement of the need for a revolution in philosophy. Twentieth-century representatives of this view include Heidegger and Wittgenstein; for instance, the latter claimed that "he was writing for people who would think in a different way, breathe a different air of life, from that of present-day men" (Malcolm 1958). It's a move that has long been characteristic of philosophy. Thus Descartes, in the *Discourse*: "Of philosophy I will say nothing, except that when I saw that it had been cultivated for many ages by the most distinguished men, and that yet there is not a single matter within its sphere which is not still in dispute, and nothing, therefore, which is above doubt." Philosophy is always in crisis, to be saved by the work of the next philosopher.

A second set of concerns is more sociological in focus, as philosophers worry about the status of the field. Jason Stanley's 'The Crisis of Philosophy', a 2010 piece in *Inside Higher Education*, expresses dismay at how philosophers are treated by fellow humanists, and complains about the lack of public recognition, as reflected in indices such as MacArthur 'genius' grants. Essays posted on *The Philosophy Smoker* express alarm about the dismal state of the job market or the lack of attention philosophy gives to race and gender issues. The *Daily Nous* puts a more positive spin on these apprehensions with its "value of philosophy pages," a collection of data and narratives supporting the benefits of studying philosophy.

A third set of apprehensions gets closer to our concerns – apocalyptic prognoses concerning the industry-wide collapse of higher education as well as criticisms of philosophy in particular. Nathan Harden (2012), noting the disruptive nature of online education, imagines that "in fifty years, if not much sooner, half of the roughly 4,500 colleges and universities now operating in the United States will have ceased to exist." Elite and well-endowed schools, the Stanfords and Bowdoins of the world, will do fine. But the mass of non-brand name schools is threatened by online education. For who is going to want a degree from a directional university when they can get an online education from MIT or Harvard at a fraction of the cost?

Then there are the attacks on philosophy in particular: scientists who argue that philosophy has been superseded by science, or politicians who claim that philosophy is a waste of taxpayer's money. Physicists can be particularly scornful: Neil deGrasse Tyson describes philosophy as something that "can really mess you up"; Freeman Dyson calls philosophers "historically insignificant." On the political side, North Carolina governor Patrick McCrory has criticized the idea that state universities should even be training PhDs in philosophy (Kiley 2013). Florida governor Rick Scott has called on public universities to cut programmes generating students "with low job prospects and earning potential." U.S. Senator and presidential candidate Marco Rubio brought the deficiencies of a philosophical education into the 2016 presidential race with his comment that "we need more welders and less philosophers." Attacks like these have put philosophy programmes on the defensive, even though it's not clear their graduates do worse jobwise than majors from other departments. These criticisms are then matched by equally pro forma defences of philosophy, as providing mental training, skills useful in the workplace, preparation for law school, etc.

Others have extended the argument to related humanities:

Within a few decades, contemporary literature departments (e.g., English) will be largely extinct – they'll be as large and vibrant as Classics departments are today, which is to say, not very active at all. Only wealthy institutions will be able to afford the luxury of faculty devoted to studying written and printed text. Communications, rhetoric/composition, and media studies will take English's place. (Pulizzi 2014)

Similarly, philosophy departments could be reduced to a rump of critical thinking courses, or even supplanted by *au courant* offerings in science and technology studies. The latter field can at least make an argument concerning its relevance in a technoscientific age – although here too what started out as a promising transdisciplinary development has increasingly suffered from disciplinary capture (cf. Jasanoff 2010).

Finally, it is worth noting that not everyone agrees that there is a crisis to address. Mark Garrett Cooper and John Marx, in a 2014 essay titled 'Crisis, Crisis, Crisis', see humanists as "champion complainers" since the late nineteenth century, when they opposed the increasingly utilitarian and scientific cast of American colleges and universities. Cooper and Marx mock the entire 'crisis' trope:

> In demonstrating that there has been nothing but crisis in the humanities for at least a century, we also aim to show that there is no crisis. ... We are bored with bumper-sticker versions of the humanities and tired of the posture of the beleaguered humanist.

Focusing on English and film studies, Cooper and Marx offer an alternative account of the humanities where there has always been a sizeable humanities workforce, even if unnamed as such. They may have emphasized their distance from the workaday world, but humanists have existed in symbiotic exchange with any number of industries – whether it was Fitzgerald and Hemingway in 1930s Hollywood, English majors in advertising firms in the 1980s, or students in cultural studies working for video game makers today. What goes by 'crisis' is simply a marker of the dynamic nature of these relations, as the humanities-business nexus evolves over time through a process that exemplifies Schumpeter's creative destruction.

VIEWS OF THE PROFESSION

Philosophy often plays a marginal role in defences of the humanities; literature is usually seen as the metonymic standard-bearer. In *The Humanities and the Dream of America* (2011), Geoffrey Harpham, then director of the National Humanities Center, tells the history of the humanities in terms of America, Harvard, and comparative literature, the latter as the successor discipline to philology. Plato and Socrates give way to Isocrates, with his focus on rhetoric rather than philosophy, and the humanities are defined as "the scholarly study of documents and artifacts produced by human beings." In Harpham's view,

> the humanities have "the text" as their object, humanity as their subject, and self-understanding as their goal. ... Other disciplines offer knowledge about things; the humanities offer knowledge about human beings, and thus imply a promise of something more than information.

The humanities focus on the cultivation of the human spirit as expressed through noteworthy texts – rather than, say, through critiques of the social, political, scientific, and economic order. Harpham's position is reminiscent of Derrida, when the latter claims that there is nothing outside the text. Critique becomes primarily an aesthetic sensibility rather than a worldly engagement.

Accounts of the humanities, offered by individuals from particular disciplinary backgrounds, habitually practice the elision of synecdoche: each assumes that their field stands for the whole of the humanities. In the modern university, the two main claimants have been literature and philosophy – the classics now diminished, art history and music classified with the arts, and history often classified as a social science. On the side of literature, it is true that English professors greatly outnumber professors of philosophy, and are heirs to a tradition extending back to Plato's *Cratylus*. But the central point for the account presented here is how the disciplinary structure of the research university evolved out of – or better said, displaced – philosophy. In the nineteenth-century American college an education in Greek, Latin, mathematics, and rhetoric was completed by what was often a yearlong course in moral philosophy, taught by the college president. It's a role that ended as philosophy became just another discipline alongside the rest.

While it in some ways continues the tradition of philology, English language and literature departments, as Gerald Graff has noted, were also part of the invention of the research university; the first departments date from the 1880s. Graff's book – *Professing Literature: An Institutional History* (first published in 1987, reissued in 2007) – anticipates several of our themes. His account of what he calls the field coverage model is essential for understanding the sociology of academic life. Under the field coverage model a department is constituted in terms of fields and topics. As Graff notes, the great advantage of this approach was to make departments essentially self-regulating. It amounts to a mutual non-aggression treaty among the parties involved: I won't challenge how you teach Milton, and you won't dispute how I teach Victorian literature. This allows departments to function not so much as corporate entities but as collections of disparate individuals who come together on occasion for departmental functions. It is a system constructed to preclude anything more than logistical questions concerning its own institutional structure.

The result has its advantages. For example, the field coverage model makes it comparatively easy to add a new field or topic (e.g. feminism or environmental philosophy) to the departmental mix. This makes it possible to imagine that someday a department might hire someone as a field philosopher. But it also implies that reorganizing a department as a whole, towards the end of creating a PhD programme with markedly different goals and approaches, is nearly impossible. The one thing held in common among all faculty members is their individual self-interest – reason enough to band together to oppose the installation of any marked departure from the disciplinary status quo. This highlights the importance to our argument that external circumstances may compel changes in institutional structures that would not occur from internal pressures.

The field coverage model also helps describe the way the sciences arose from philosophy. The first recorded use of the term 'natural science', by William Whewell, occurred in 1833; before that, the portmanteau term was natural philosophy. The physical sciences quickly developed in the 1860s and 1870s as the modern research university acquired majors, departments, and specialists. About the same time the seven social sciences evolved out of moral philosophy. The natural and social sciences now divided the intellectual world between them, as a set of regional ontologies. The role that philosophy had played in looking at knowledge as a whole was undercut by Darwin's destruction of natural theology: there no longer *was* a whole, at least in the sense of knowledge adding up to a larger meaning that united the natural and social spheres. That is, the death of natural philosophy – as contrasted with environmental ethics, which today is largely the domain of specialists with limited interests – marked the end of the hegemony of philosophy.

These institutional shifts posed the question of what would happen to philosophy. Some thought the field was an anachronism destined to pass from the scene. Instead, having lost much of its empirical focus to the social sciences and its connection to nature through the rise of the physical sciences and the death of the argument from design, philosophy became another regional ontology largely concerned with logic and the philosophy of science.

Of course, there are philosophers who are happy with the status quo, who see only clear skies; who view the welter of problems facing the discipline as lying on the periphery of the profession. Yes, funding is down, at least in the provinces, and the graduate students from lesser programmes face a poor prognosis in terms of employment. But this is merely a sad case of the survival of the fittest. Teaching loads may become onerous because of purblind administrators and legislators, although this too is mainly a problem for the weaker schools. But theorizing, the beating heart of philosophy, has never been healthier. We are living in a golden age of philosophy – the subtlety of contemporary philosophic work unmatched, the opportunities for publishing vast, the quality of graduate students unprecedented. Thus,

> unlike many other disciplines in the humanities and social sciences, which in recent years were seduced by bad French philosophy into a lot of silly "post-modern" theorizing that exposed them to derision and reduced them to irrelevance, analytic philosophy is flourishing. Part of the reason why analytic philosophy generally is in such a healthy state is that, as Jerry Fodor observed in a recent book review, philosophers no longer tend to have philosophies. We no longer devote our lives to developing comprehensive philosophical or ethical systems. (McMahan 2009)

Sanguine views like this are helped along by the sinecure of tenure, while at the same time revealing the class structure of the discipline.

Some dissenters at prominent schools get closer to our view. In his 2014 book *Empty Ideas: A Critique of Analytic Philosophy*, Peter Unger of New York University disparaged the pretentions of contemporary analytic philosophy. Philip Kitcher of Columbia University has compared contemporary philosophy to musicians who spend their time "adding an extra trill to Quadruple Tremolo 41 ... [of interest to] a tiny group of self-described cognoscenti" (Kitcher 2011). But these authors treat the problems of the discipline as a matter of personal predilections or intellectual trends. There is no hint that the problem might be institutional in nature, the reciprocal effects of university structures and patterns of theorizing. Similarly, University of Montana philosopher Albert Borgmann writes that the biggest problem with philosophers is what they are *not* doing: "It is what they are leaving undone that is so troubling, their failure to recognize and overcome their seclusion or exclusion from the public conversation" (Borgmann 1995, p. 304). But he doesn't spell out what this means. *How* would philosophers go about playing a greater role in the public conversation? Writing books and articles primarily addressed to one's disciplinary peers is a pretty indirect way to go about it. Borgmann is silent on the question of alternative pathways to impact.

Bruce Kuklick (2001) comes closest to our assessment by framing the matter as a problem of audience. He points out that "ensconced in the university system, a discipline can exist for a long time with a minimal audience, although even the leadership of the institutions of higher education may catch on after a time." Since Kuklick's assessment, these political imperatives have only become more pressing. How much longer will administrators – and state legislatures – provide financial support for small groups of scholars to speak only to one another?

Nonetheless, philosophers and humanists devote little attention to the issue, offering homilies instead of developing approaches responsive to contemporary realities. Where is the graduate course that thematizes the question of how philosophy could specifically contribute to the work of scientists? Where do we find accounts of philosophers or other humanists offering critiques of the metrics being developed for evaluating academic work – or working up alternative accounts for defining excellence? A few humanists thinking are about these matters (e.g. Newfield 2011), but not nearly enough.

A further sign of inattention is the absence of a grey literature of studies and reports on the future of philosophy. The American Philosophical Association (APA) website contains nothing that qualifies as an analysis of how the profession could better respond to changing economic, political, and cultural trends. Now, one does find statements about the importance of noting that 'history of philosophy' classes are typically classes in the history of *Western* philosophy. And there is a set of informational guides that survey topics such as 'A Non-Academic Career?' prepared by the APA Committee on Non-Academic Careers – although these date from 1999, "with a few updates in 2002." The APA also publishes an annual 'Guide to Graduate Programs in Philosophy'. But this merely consists of summary compilations of admission data from departments (e.g. student enrolments and entry requirements).[1] One finds no white papers on the status of the profession. This stands in marked contrast to the Modern Language Association (MLA). The MLA has convened groups of scholars and published a number of reports on the status of the English profession that seek to understand how English can respond to changing social dynamics (e.g. MLA 2014). These analyses usually stay at the level of economic concerns (how do we get jobs for our graduate students?) rather than raising questions about the raison d'etre of the modern research enterprise itself. But they are at least a step in the right direction.

It all adds up to a non sequitur: for a discipline that has been concerned with questions of temporality and history at least since Hegel, academic philosophy operates in a bizarrely ahistorical fashion in terms of its institutional embeddedness. Philosophy of science courses overwhelmingly outnumber philosophy of technology courses, even though students today have intimate daily contact with the latter rather than the former. Logic courses are still text based rather than exploring the logic of images and video. The perennial questions of philosophy will never pass from the scene – what would philosophy be without Plato? But without the cultivation of a more innovative and entrepreneurial academic culture, philosophical research (and humanities research generally) is likely to become more and more circumscribed to a small number of prestigious institutions that can afford the luxury of producing work of interest to only a handful of specialists.

In sum: for all the tumult surrounding 'crisis' there have been few *philosophical* attempts to understand the current situation of philosophy. This point is liable to be misunderstood, so let us be clear. Of course there is an abundant literature in the history of philosophy on how philosophical ideas have affected society and are embedded within a social context. And philosophers are constantly making attempts (most obviously in applied philosophy, as well as in a number of other subfields) at shedding philosophical light on contemporary problems of one sort or another. But these accounts, worthy as they are, are *entirely* different from our point here: understanding how the current institutional structure has handicapped philosophy from playing a more vibrant role in society. Philosophers make de rigueur references to the vexed relationship between the philosopher and the polis, and cite Socrates as the classic example. But there are few accounts of attempts to embed philosophy in particular contexts, describing either success or failure; or of philosophers working alongside scientists and policymakers, or engineers and marketers, in the public or private sectors; or of graduate students interning at the NSF or the Environmental Protection Agency. This simply is not considered to be 'philosophical'.

VITAL SIGNS

Crisis or not, some will view the whole matter as a tempest in a teapot. As of 2013, the American Academy of Arts and Sciences found that liberal arts majors accounted for just 7% of all majors. Philosophy majors comprise one half of 1% of the total. If there is a crisis, it concerns only a small number of people. In the United States, there are only some 24,000 philosophy faculty across post-secondary education.[2]

But this reduces philosophy to its physical footprint. Even today, with little explicit thought about impact, this group plays an outsized role in society. Perhaps the best (if perhaps not the most saleable) defence of philosophy is the simplest: society needs a class of people with the time and, yes, leisure to think through questions concerning the meaning of life, expanding our moral imagination and challenging shibboleths, even at the risk of foolishness or censure. The results of this work already seep out to society, through the long-echoing impacts of teachers in the classroom and the slow cultivation of scholarship. It just would happen more often and directly through the more systematic set of interventions like those that we are advocating here.

It's difficult to get a grasp of the vital signs of academic philosophy. Good numbers are hard to come by, precisely because of the lack of institutional awareness that we are highlighting. To gather data, we sometimes have had to look at trends pertaining to the humanities or the academy more generally.

Or we have had to conduct our own surveys – what might count as 'bad social science', perhaps, but which we hope will spur others to improve upon our work.

In what follows, we survey indicators of philosophy in the United States across the five dimensions that are most often referenced in crisis narratives and reports about the state of academia: (a) enrolment and degrees awarded; (b) faculty workforce; (c) academic job prospects for PhD graduates (and undergrad major earnings); (d) scholarly production; and (e) revenue/costs (encompassing tuition, state support, and student debt).

(a) Enrolment. Commentators have argued that the humanities are in trouble: both enrolments and degrees conferred have dropped, as students jump ship for business and the STEM fields. *New York Times* columnist David Brooks (2013), for example, argued that fifty years ago there were twice as many humanities majors as there are now (14% rather than 7% of the total). On the other hand, drawing from several data sources, Michael Bérubé and Jennifer Ruth (2015) have argued that the 'crisis' of enrolment is a myth, and that the percentage of humanities majors has held steady since the 1970s. Ben Schmidt (2013) notes "we give out far more population-normalized degrees in the humanities now than we did in the 1950s or the 1980s." Data from humanitiesindicators.org show bachelor's degrees awarded annually in philosophy have more than doubled in the last twenty-five years, from 3,534 in 1987 to 7,842 in 2013 – a higher rate of growth than that of the college age population, which grew by 50% across that time.

That said, claims of an enrolment crisis in the humanities largely turn on how one frames the issue. The danger lies in comparing apples and oranges, making assessments across the changing landscape of higher education. The increase in community college students skews comparisons, as does the growing enrolment of part-time students and changes in their gender distribution. Moreover, it wasn't possible to major in fields like computer science fifty years ago. The explosion of knowledge, the development of specialties, and the process of democratization of higher education have all affected the distribution of majors. In the face of this, numbers seem to have increased not just in the United States but also in the United Kingdom and Australia (see Mandler 2015).

(b) Faculty workforce. Bérubé and Ruth argue that talk about enrolment distracts from the real crisis in the humanities, which is "not one of disappearing students. It is one of disappearing tenure-track jobs" (p. 10). Thus, Keith Hoeller (2014) speaks of the "Wal-Mart-ization of higher education," noting that in the past thirty-eight years, the percentage of professors holding tenure-track positions has been cut nearly in half. (Once again, numbers vary here in part because of different criteria: whether to include courses taught by graduate students – in our own department, fully 50% of the courses are so taught – or whether to limit the discussion to tenure-track positions, lecturers, and adjuncts.) Counting all types, full-time tenure-stream professors went from 45% of America's professoriate in 1975 to 24% in 2011. Viewed from another perspective, from 1975 to 2011, in terms of absolute numbers, the number of tenure-track and tenured professors increased by 35.6% nationwide, while the number of part-time (or contingent) professors increased by 325% (Figure 2.1).

Hoeller uses Wal-Mart as an analogy because of its policy of keeping full-time workers to a minimum, while hiring part-time workers with low pay, no benefits, and no job security. In the academy, this is problematic not just in terms of fair pay – as Bérubé and Ruth note, it's telling that colleges are considered a pathway to the middle class even as they pay their non-tenure track instructors "food-stamp wages." Equally troubling is the erosion of faculty self-governance and academic freedom, as administrators are able to more effectively apply pressure to a vulnerable, contingent academic workforce.

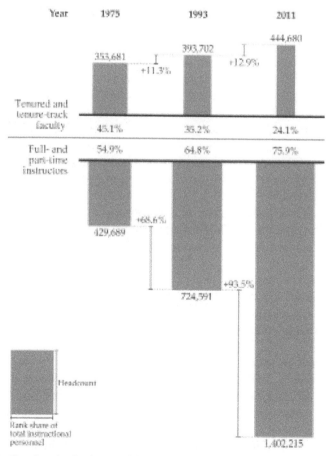

Figure 2.1   The changing landscape of the faculty workforce. *Source*: AAUP.

Declines in autonomy thus undermine the university's core mission of seeking knowledge and fostering critical thinking in the service of the common good.

(c) Jobs. Labour market dynamics clearly bear on a third indicator – the academic job market. Data here is spotty: the APA, for instance, puts little effort into collecting placement statistics. The account we cobbled together from the APA, including historical Jobs for Philosophers data, does not even distinguish graduates from PhD programmes as compared to seminaries. Nor does it distinguish the kinds of jobs listed in any given year – there is no breakdown between tenure track, adjunct, and lecturer positions, or any attention to trends over time.

There has been one attempt to systematically survey the philosophy job market (Carson 2013). It shows that placement statistics are good at the 'top rated' departments (as ranked by *The Philosophical Gourmet*), with for example, 91% of PhD graduates from Yale's philosophy programme currently holding a tenured or tenure-track position. By the time you get to the 35th-ranked school, the odds of a PhD graduate holding a tenured or tenure-track position drop to 50%, and drop again to 13% for the 60th-ranked PhD programme. Another forty or so programmes, including our own at the University of North Texas, do not even make the list.[3]

Another study looked at the 2015 positions listed on the site 'Philjobs' and broke them down by type (Arvan 2015). That analysis turned up 152 advertisements for tenure-track jobs and 97 for non-tenure-track jobs. In both cases, the areas of specialization with the most openings were ethics, applied ethics, and philosophy of science. Indeed, among tenure-track jobs, 35 were in ethics or applied ethics, while only one was in philosophy of language and one in metaphysics.

There is also not much data on the job prospects of undergraduate philosophy majors. However, the site PayScale.com does track the income of 1.4 million college alumni and breaks down career earnings by major. Of the 319 majors tracked, philosophy majors rank 75th in career earnings (see Lam 2015). This makes it the top ranked humanities degree and even puts philosophy majors above degrees such as information technology, accounting and finance, and business and marketing. Philosophy majors can expect over $650,000 extra in lifetime earnings compared to other humanities majors (see Dorfman 2014).

(d) Scholarly production. Consider first overall scholarly productivity: it is estimated that in 2006 there were 1.35 million peer-reviewed articles published (Björk 2008). Derek de Solla Price (1961) estimated that the number of scholarly journals had gone from one in 1662 (the *Philosophical Transactions* of the Royal Society) to 60,000 in 1950. He extrapolated the annual growth rate (5.6%, or a doubling every fifteen years) to predict there would be 1 million journals by the year 2000. Indeed in 2002, there were 905,090 periodicals that had been assigned ISSN numbers. A later analysis (by Jacobs 2013) argued that of those, about 28,000 truly count as peer-reviewed journals. They also revised Price's growth rate estimates downwards, but in their own analysis of several databases from 1907 to 2007, they still found growth rates around 4% and a doubling times around eighteen years. This increase in production mirrors the rise in total PhDs awarded (see Figure 2.2).

Bruce Kuklick (2001) notes that philosophy has followed this general pattern of rapid growth:

> In 1920 the membership of the American Philosophical Association was about 260; in 1960 it was 1,500; in the 1990s it was well over 8,000. One observer noted that in the first half of the twentieth century the United States, Britain, and Canada founded thirty philosophy journals. Fifteen more were added between 1950 and 1960, and forty-four in the 1960s – as many as in the previous sixty years – and then about 120 in the next twenty years!

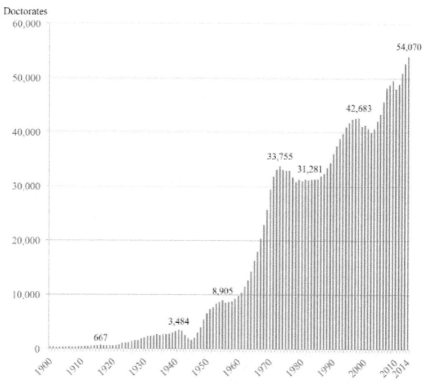

Figure 2.2 Total number of awarded doctorates annually in the U.S. across all fields, 1900–2014. *Source*: Figure adapted from Thurgood, L., M. J. Golladay, and S. T. Hill, "U.S. Doctorates in the 20th Century: Special Report" (NSF, June 2006). Data sources: NSF/NIH/USED/NEH/USDA/NASA, Survey of Earned Doctorates and Doctorate Records File (1920–2014) and U.S. Office of Education annual reports (1900–1919).

In 2015, PhilPapers, a comprehensive index of philosophy journals, listed 1,032 titles in its collection.

Is this a problem? The over-production of disciplinary scholarship is not like the over-production of goods in a market. In a commodity market, if goods are not consumed, then prices sink as supplies soar. This leads to devastating (and corrective) impacts for the producers. But in the disciplinary economy, even if no one consumes (i.e. reads) the products, they do not lose value – at least in terms of getting their producer another line on his or her CV, and perhaps tenure or promotion. In other markets this situation would culminate in drastic contraction. But the production of knowledge has just kept chugging along. And it will, unless an outside factor steps in to change the dynamics. If state legislators look into these numbers, and the corresponding lack of citations for scholarly output, we may see a crash driven by an increase in teaching loads that preclude research.

Some have argued for adjusting the ways in which faculty are evaluated by decentring the peer-reviewed publication as the coin of the realm (cf. Wittkower, Selinger, and Rush 2014). Such changes, however, still seem to be a very long way off. A look at the criteria for tenure decisions utilized by the 750 philosophy departments shows that only 4% value "public humanities" (see Figure 2.3). In our own 2010 survey we found that extra-disciplinary efforts, that is, the ability to attract funding, engage in applied research, or publish outside of philosophy journals continued to rank at the bottom of the criteria for tenure and promotion (Frodeman 2011).

(e) Revenue and costs. A 2014 Government Accounting Office report describes the situation in stark terms: "From fiscal years 2003 through 2012, state funding for all public colleges decreased, while tuition rose. Specifically, state funding decreased by 12 percent overall while median tuition rose 55 percent across all public colleges." Over that same time period, the costs of college grew nearly twice as fast as healthcare costs. Nationally, the cost (in constant dollars) of attending a four-year public university more than tripled from 1984 to 2014 (College Board 2014). At some places it was worse than that. In the University of California system, for example, in-state tuition was $300 in 1980. By 2014, it was more than $11,000. By 2014, the University of Colorado had nearly completed a transition to a private institution, drawing just 5% of its funds from the state.

Bérubé and Ruth describe this as an act of intergenerational betrayal: the baby boomers who benefited from subsidized higher education turn around and defund it once they are in power. The year 2012 marked the first time tuition revenue paid by students surpassed the amount of funding states provided for public universities (GAO 2014). Higher education is increasingly privatized and geared towards job preparation, a trend that calls into question

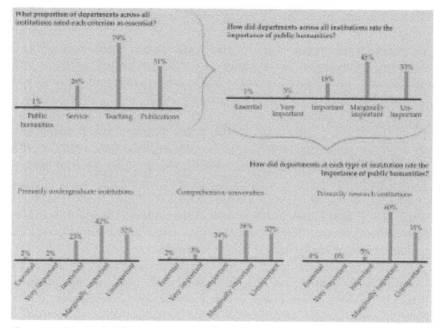

Figure 2.3   **Tenure decision criteria in US philosophy departments, 2007–2012.** *Source*: Data from humanitiesindicators.org.

the roles of universities to provide an education in our culture's history, training in democratic principles, and fostering citizenship.

These trends have saddled students with increasing amounts of debt. Current student debt load is estimated at $34,000 per household or roughly $1 trillion total, three times as high as it was in 1989. While no one is celebrating this, the extent to which it is a problem is unclear (see Allan and Thompson 2013). For example, some argue that the payment to income ratio over that time has fallen, meaning that there is actually less of a debt burden now (Akers and Chingos 2014; though see Weiss 2015). The student debt picture may not be as grim as it is often made out to be, thanks in large part to recent U.S. federal policies including income-based loan repayments, caps on monthly repayment obligations, and expanded loan forgiveness (see Delbanco 2015). One author even concludes that we are at the start of a "quiet revolution in helping lift the burden of student debt" (Carey 2015).

THE NEED TO CHANGE

The numbers tell a complex and ambivalent tale. Nonetheless, we can distil an overall account of the situation. Student interest in philosophy and the humanities remains steady (if not growing) at both the undergraduate and graduate levels, while faculty autonomy is being eroded as tenure-track appointments disappear. Adjuncts and graduate students constitute an increasing percentage of the teaching pool – the former lacking decent salaries, benefits, or security; the latter drawn into graduate programmes with poor hopes of finding a tenure-track position – with the exception of those at top programmes. Many graduate programmes find their existence underwritten by the need for a cheap labour pool to teach lower division undergraduate classes – never mind the long-term prospects of employment for PhDs in philosophy. Meanwhile, research production continues to rise, even as society begins to question whether this scholarship contributes to anything or does any good. Increasingly guided by neoliberal assumptions, society is suspicious of humanities research, instead focusing on utilitarian metrics of returns on investment. Society seems intent on applying market mechanisms to domains traditionally insulated from the market.

What future does this portend? Few appreciate how all the pieces of the current system stand or fall together. One cannot eliminate graduate programmes without losing cheap labour for undergraduate courses, as well as enrolments for graduate courses where professors teach their research specialty. Lacking graduate students, professors would likely face markedly higher teaching loads. A three-tier system could emerge: a small number of comfortable and well paid tenure-stream faculty at prestigious places; the bulk of professors teaching introductory level courses with little time for research; and an even larger cohort of dispensable lecturers and adjuncts, the latter teaching courses at $3,000 each – with no benefits or job security. And this is not to even explore the possibilities of what might result from the migration of philosophy courses online. Note finally that while such a system might seem to leave the top twenty programmes in fine shape, where do these schools place their students if the bottom half or two thirds of philosophy programmes disappear?

While the future is uncertain the trend lines are clear. These vectors point towards a likely choice when it comes to philosophical research: prepare for a markedly diminished future. Or: create new models for philosophy that open up new opportunities for research and employment.

Yet there is not a single PhD programme among the 110 or so in the United States that has explicitly tried to respond to these factors. One can find a few programmes that highlight a commitment to applied philosophy, a point we will explore in the next chapter. But there is nothing like a concerted institutional effort to devise new models for what counts as philosophical research. Indeed, the very idea is considered risible: there is only good and bad philosophy, serious and amateurish attempts at the one (disciplinary) model, as judged by the professionals within the field. Emblematic of this is the one-dimensional scale of quality of the Philosophical Gourmet Report, which is somehow able to parse philosophical excellence on a single numerical scale (measured to the tenths of the point, no less; for a critical reading of the PGR see Pedersen 2012). Nor are there plans by the APA or others to better track the basic indices of the profession, or to treat the future of the profession as itself a matter worth serious thought.

Why so little action when the writing is on the wall? The problem is basically sociological in origin. Upton Sinclair made the point in *The Goose-Step: A Study of American Education* (1923) in a chapter called 'The Professor's Union', where he notes that "the anarchist attitude goes with the intellectual life." Each professor was once the brightest kid in class; each is confident of their individual superiority, and has made a career of going his or her own way. Class-consciousness implies a group solidarity that does not come naturally to the professoriate. It is often said that academic politics are so vicious because the stakes are so small. But this is wrong: academic politics are often vicious because people have little incentive to act politically – to compromise and work together to build towards a common good. It's a predictable effect of tenure, as well as the result of the field coverage model where everyone has their own period or field, or rather sub-sub specialty, where everyone can be the king or queen of that domain.

Philosophers will find it increasingly difficult to convince the state legislature to continue to subsidize the umpteenth exegesis on Kant. And as much as we are committed to philosophy, we can't really blame non-academics for their scepticism. Philosophical research has been (mis)conceived along the lines of science – even today philosophers speak of the need to be *wissenschaftlich*. But philosophy should not be, or at least not solely, another silo of specialized knowledge alongside the others. Philosophy's nature is to be dendritic, undomesticated, and cross-cutting – getting its hands dirty, changing the world and itself in the bargain.

NOTES

1. http://c.ymcdn.com/sites/www.apaonline.org/resource/resmgr/Grad_guide/gg14_complete.pdf. Noelle McAfee, at her blog "gonepublic: philosophy, politics, & public life" – has a number of posts on the need for a better database on the state of the profession.

2. Numbers from humanitiesindicators.org. The graduate students, of course, are at the 100 or so graduate programmes in philosophy in North America.

3. Job placements from our relatively young (2005) PhD programme show UNT as placing about 25% of graduates in tenure-track appointments.

Interlude 1

Philosophical Places

PHILOSOPHY ABOUNDS

Humans are thinking creatures and when we are thinking, even dimly, the world can appear as a mystery. Questions open up and are ripe for wondering. Now, many questions are trivial, or perhaps they have already been solved and it's just a matter of getting hold of the answer, or maybe they are not yet solved but there is a method for doing so. But there are other questions that call for reflection about what is or what should be. Thinking about these questions – wherever they may arise, in personal or public life – is philosophy.

Philosophy is not like other fields. Philosophy is pervasive in a way that chemistry and biology are not. Yes, we are in some sense chemical and biological beings. But we don't have to think through chemistry or biology in the same way we have to think through philosophy. We can just let metabolism happen but we can't just let thinking happen. We have to do it.

Philosophy is often interstitial, popping up here and there in odd places and inopportune times. Think of the conflicting desires and obligations in your personal and civic life. How to spend time more wisely? How to treat her? What to say to him? When to let someone fail and when to intercede? When to respond playfully and when seriously? In public life, what do individuals owe one another when it comes to healthcare, humanitarian aid, and social security? If beauty is subjective then why do we flock to national parks? When talking with others, when do we make an argument, versus responding with a laugh or with silence?

The only way to avoid such entanglements is to lead a thoughtless (and thus less than fully human) existence. This is to take states of affairs and your role in them as simply given. In this way, Eichmann kept the trains running.

Less dramatically, this also marks the passivity of too many evenings spent watching television or playing video games. Both cases denote flights from the opportunities and responsibilities of life. We have to interrogate the world if it is going to show itself in its splendour. We are called to expand our moral imaginations. Machines do not question why or wherefore. But for humans to behave like a machine diminishes their lives.

Whoever thinks about such questions is a philosopher. We are not all engineers or computer programmers, but we all are called on to be philosophers. Philosophy is a fundamental rather than a regional ontology – which is why departments of philosophy are such a poor idea. But the ubiquity of philosophy doesn't mean that everyone is equally skilled at such thinking. There may be no 'right' answer to philosophical questions as there is with arithmetic, but there are more or less thoughtful ones. Thinking out such questions is a difficult task. Academic philosophers have a leg up even when they are not the most naturally gifted thinkers: they are trained in making arguments, they are familiar with philosophical concepts and the history of ideas, and they have the time to collect evidence and do the careful thinking. Most people have other duties that call them out from this reflective space.

PHILOSOPHY ABOUNDS, BUT IT IS AFFORDED NO SPACE

If philosophy is present everywhere, then why do academic philosophers play such a marginal role in society?

One reason is that the issues troubling society are not seen as philosophical. In the wider world, the philosophical dimensions of a problem rarely come in a pure form; they lurk within other issues. This makes them easy to ignore. We look for economic and scientific solutions to our problems – answers that do not call upon our inner resources. (It is thus no surprise that spending for humanities research is just one half of 1% of the amount dedicated to science and engineering R&D.[1]) Technical experts define the nature of problems without anyone noticing the legerdemain whereby ethical, aesthetic, and metaphysical issues have been turned into technical ones. And so the field of economics, once known as political economy, armours itself with equations that obscure the fundamentally philosophical nature of many of the questions it addresses.

Science, technology, and economics function as escapes from the open-ended questioning of philosophy. They are seen as value-neutral tools for action. This leads people say the most remarkable things. "What does philosophy have to do with fracking?" Meanwhile they are wrangling about matters of risks, rights, and responsibilities. Or "what does philosophy have to do with climate change?" Meanwhile they worry about droughts and snowpacks and a feeling that something special has been lost from their lives. Philosophers are marginal, but not because philosophy isn't present. It's because no one sees the philosophy and thus no one thinks to invite a philosopher to the table.

To understand what kind of society we have – or ought to have – one must consider the purpose of institutions and the kinds of honours that ought to be bestowed. Such thinking can be called first-order – if this is understood as first in importance rather than in time. Philosophy is fundamental, but it is often cryptic. You don't first 'get the philosophy right' and then apply it or move on to the non-philosophical stuff. Questions of meaning and purpose mostly lie below the surface, liable to pop up at any moment.

Foregrounding philosophy, though, invites problems. Reasonable disagreements about the meaning of human flourishing can degenerate into warring dogmatisms. People grow insecure, and give in to the itch for certainty. Insights turn into dogmas. Your livelihood can be imperilled if you do not adhere to a certain comprehensive doctrine, whether secular or theological. Realizing this, theorists like Hobbes and Locke found reasons for marginalizing questions of the good. Thus classical liberalism was born. Whenever possible, turn political questions into technical issues; otherwise let people do as they please.

The attempt to privatize matters of the good – questions of goals and purposes – was an effort to open the relief valve on the pressure cooker of society. You can have your own personal 'philosophy of life', just don't foist it on me and we'll get along fine. Increasingly, however, this approach shows itself as dysfunctional, as private preferences create public realities for everyone. Thus we can hardly choose to not have a cell phone. Such choices also hold hidden commitments. So we agreed to the common goal of economic growth, because we thought of money as mere means – we can purchase what we want with it. But the money form contained its own values: defining our lives in terms of 'having' rather than 'being', and living a life increasingly distracted by shiny objects.

The question of the good can be turned into a private matter only under specific conditions. We mean ecological abundance: there must be sufficient space and resources for people to do their own thing (a New World helps). Both Locke's political philosophy and Mill's ethics are predicated upon abundance. Now, on an increasingly limited planet, we have little choice but to again 'publicize' the question of the good. And to do philosophy in a public way. Latour (1991) was correct: we have never been modern. We have never truly privatized the question of the good.

EVEN WHEN PUBLIC SPACE IS OPEN,
PHILOSOPHERS RARELY ENTER

But philosophers are not only victims of a shrunken public space. Even when an opportunity arises to make a contribution, philosophers are unlikely to, as Hannah Arendt puts it, "go visiting." Indeed, instead of protesting their fringe status, philosophers have too often treated it as a refuge and virtue. Too many philosophers have agreed with Strauss: "the selfish or class interest of the philosophers consists in being left alone" (Strauss 1958, p. 142).

Academic philosophers have a collective identity as members of a community, which is bestowed not so much by the things they discuss as by the institutions that house and credential them. The PhD allows for the distinction to be made between the real philosophers and the amateurs. This certification affords a measure of power in the form of credibility. It gives philosophers a platform so that their voice can be heard above the din of the masses. Ideally – and this should be the point of tenure – one can speak from a place of conviction concerning the common good. Authoritative if also fallible, philosophers could function as both intellectual foil and moral conscience of society.

It hasn't worked out that way. Philosophers rarely play a public role. Bruce Kuklick argues this is because the dominant strain of analytic philosophy classified moral and political issues as outside of the cognitive domain – mere opinionating, and thus not part of the proper remit of the philosopher. The relativism of many twentieth-century philosophical camps – stemming from the likes of Ruth Benedict and Thomas Kuhn – gives them a similar aversion to public engagement. It's hard to claim an authoritative role in society when you are sceptical of all claims to privileged knowledge, including your own. And so sociologists track indicators of 'subjective well-being' because identifying one way of life as better than another is considered a faux pas.

Nietzsche had a different view of the philosopher – as someone who "demands of himself a judgment, a Yes or No, not about the sciences but about life and the values of life" (*Beyond Good and Evil*, 205). This attitude is out of step with our intellectual culture, which resists authoritative claims, instead swinging from scepticism to certainty and back again. The philosopher, however, can still offer a distinctive voice. That voice manifests itself differently than in the sciences. The sciences, natural and social, are in the business of providing answers. But answers aren't the only way to contribute to a conversation. Socrates' professions of 'ignorance' actually show the philosopher's skills of probing, questioning, and unsettling matters: expanding our moral imagination. Tone is crucial here, for questioning can be destructive. But when well wrought, questioning opens up a new view on an issue, rendering problematic what was previously thought to be given.

On our reading, the challenges of contemporary society are now more philosophical than technical in nature. But we face two intertwined problems: the flight of society from philosophy and of philosophy from society. We will focus more on the second problem, but the first also deserves careful thought.

NOTE

1. From the Humanities Indicators, 'The State of the Humanities: Funding 2014' report, p. 4: http://www.humanitiesindicators.org/binaries/pdf/HI_FundingReport2014.pdf.

Part II

DISCIPLINARITY AND ITS DISCONTENTS

Chapter 3

Applied Philosophy

Being useful doesn't come easy for philosophers. Some reasons for this are perennial. The chief one is that philosophers are liable to question the meaning and value of utility. A little perversely, perhaps, we argue that such questioning is in fact rather useful. After all, we want things to be useful *for* something. We don't want the wrong things to be useful, or to grease the wheels as we skid in the wrong direction.

There's a standard reaction to such comments. Things are 'useful' if they bring me pleasure, and I am the first and final arbiter of what counts as pleasurable. But this is wrong, of course. Things are more complicated than that. We'll consider what we mean by 'useful' in a moment, as we turn to a consideration of applied philosophy. But it's worth noting that we are often mistaken about what brings us pleasure. We have to learn to appreciate the best pleasures, and disdain the bad; both points take attentiveness and discipline. Altogether, cultivating the art of well-being (*eudaimonia*) is not a simple project.

The useful has come to be defined in terms of changing things. It's the old Baconian definition that knowledge is power; it's what Heidegger calls our productionist metaphysics. When people say that philosophy is useless they mean it doesn't *do* anything: philosophy bakes no bread. Granted, we need bread; but we also need much more than that. And so this criticism constitutes an odd, even counter-intuitive dismissal of what Aristotle called contemplation – an experience which we are all familiar with, in the appreciation of a sunset, our child's musical performance, or in convivial conversation over dinner. The value of many of life's best moments lies in the bare fact that they *are* rather than in their producing anything.[1]

Adam Smith (of all people) scrambles our categories when he distinguishes between "productive" and "unproductive" labour in *The Wealth of Nations*

(1776). Productive labour "fixes and realizes itself in some particular subject or vendible commodity, which lasts for some time at least after that labour is past." Unproductive labour does no such thing – such services "perish in the very instant of their performance, and seldom leave any trace or value behind them for which an equal quantity of service could afterwards be procured." The tune of the musician fades as soon as it is sung (see book 2, chapter 3 for this discussion).

Though he doesn't explicitly say it, it's a safe bet that Smith categorized philosophy as unproductive labour. This puts philosophers in company with "some of the most respectable orders in the society." Judges, lawyers, soldiers, churchmen, physicians, musicians, and even the sovereign himself are all unproductive labourers. As he notes, you can't sell the service of soldiers last year to buy defence for the country this year. That service vanished in the very act of serving. But, for Smith, this does not mean that such labour is useless. Indeed, he describes certain kinds of unproductive labour in glowing terms as "honourable," "necessary," and in fact "useful."

Productive labour is useful in one sense, in that it supplies the annual "fund" that constitutes the wealth of nations, namely the "necessaries and conveniences of life." Unproductive labour draws down that fund. But that doesn't make unproductive labour useless. It is useful in another, perhaps higher sense – neither productive nor consumptive, but rather contemplative in the sense noted above. It is what all that production is *for*, once we put our restless "trucking disposition" into neutral, pausing our ambition and acquisitiveness for a while. As Smith notes, a high proportion of unproductive labourers is perhaps the chief mark of "civilized and thriving nations."

A good life, then, needs to be useful in both a productive and non-productive sense. But there's another wrinkle to the story. The musician's tune no longer perishes in the moment. Through recording devices it is now stored up as a "vendible commodity." It would seem that the categories are fluid, as unproductive labour can become productive. This happens with philosophy, too. After all, philosophers churn out books, articles, and videos (in bookstores and behind pay walls) as well as engaging in talk.

In itself, the productionist aspects of philosophy aren't ominous. Yes, they can be harbingers of a certain narrow-minded neoliberalism or utilitarianism that reduces all meaning and value to productivity. But it need not function that way. Rather, as we will argue below, the danger lies in applied philosophy's embrace of the banking notion of philosophizing, where the work has already been done and is stored up, ready to be applied somewhere down the road.

Applied philosophy and philosophy more generally have too easily fallen into a dialectic of usefulness and uselessness – with plenty of scholars defending the dignity of the latter. This is a shame, for we need an expansion of our notion of what counts as useful. Part of the problem with applied philosophy is that it has become too productive – of texts in peer-reviewed journals – neglecting the 'unproductivity' of efforts that vanish in the doing, the performance of working alongside people, helping them with their problems. It has conflated 'useful' with 'productive,' rather than seeing its usefulness as often being tied to the evanescent and unproductive labour of talk. Applied philosophy is often offered as an antidote to the irrelevance of philosophy, but it is beholden to the same dynamic that is responsible for philosophy's marginal place in society. It is a productionist enterprise, storing up great funds of knowledge, funds that are rarely drawn down by society and put to use.

Between them, Aristotle and Smith highlight the usefulness of an unproductive philosophy. But this raises a problem for a latter-day Socrates. Just how do you evaluate someone whose value-added floats off in the breeze as soon as it leaves his lips? Socrates had real effects – on the minds of the youth and the character of the polis – but his work did not register as a commodity or a product. (Indeed, Socrates was notorious for his criticism of writing.) It would help if we can find indices that could track the effects of unproductive activities (and different kinds of productive activities) on the larger world.

TWO SITES OF APPLIED PHILOSOPHY

When the NSF was established in 1950 its mandate was clear: fund basic research in math, engineering, and the physical sciences. The simple act of funding basic science was thought to equal 'science in the public interest'. No special efforts were necessary to make research relevant. The products of research would trickle out into society through a natural osmotic process. It was not until 2006 that the NSF thought of this as enough of a theoretical problem to establish a funding programme – Science of Science and Innovation Policy (SciSIP) – to study how scientific insights come to have an impact on society.

NSF support was originally limited to the natural sciences, math, and engineering. Funding for the social sciences was considered but rejected[2]; funding for the humanities was never a possibility.[3] Yet it was only a few years – 1956 – before the NSF created a small programme in the history, philosophy, and sociology of science. It turned out that doing science could not be divorced from reflection on science. In the real world, problems do not break into disciplines, and scientific work is all mixed up with political and philosophic issues. Both science and society needed help in how to think about these issues.

De facto the NSF had created a programme in applied humanities and social science, with research intended to lead to real policy outcomes. But

the assumptions underlying this work remained disciplinary in nature: support philosophers (historians, sociologists, etc.) to do high quality work; let
this work be vetted by disciplinary peers; the work would then be published
in disciplinary venues. There was no need to make a special effort to engage
outside audiences – no need to send philosophers into the lab or field, or
to have historians work in an ongoing manner with STEM researchers or
policymakers. Simply turn such work into productive labour, creating a reservoir (fund) of knowledge that could be drawn upon by anyone who was
interested. After a half-century of support the results of this approach constitute something like a database. How has this disciplinary and productionist
model worked out? What has been the impact of NSF-funded social scientific
and humanistic research on either the STEM community or society at large?

We discovered that no one had conducted a meta-study of this type. And
so in 2012 we submitted a grant application to the NSF to examine the
philosophical literature that resulted from this support, in order to see in
what ways it had been taken up by the STEM disciplines, policymakers, or
society. Our request for funding was turned down. We reapplied, and our
proposal morphed through a series of rejections and resubmissions. In 2014
we received partial funding; by that point the project had come to focus on
what can be called the *other* birth of applied philosophy, within the discipline
of philosophy itself. Our work would focus on the scholarly trends since
the 1980s rather than on the NSF-supported work dating from the 1950s.
We would search the applied philosophy literature for accounts where philosophers described their successes (and failures) in having an impact upon
groups beyond the philosophic community. By doing so, we hoped to identify
a set of best practices for how philosophers could have a broader impact.[4]

In 2014 we also received (as part of a larger team) a second and related
NSF grant to examine the ways in which the broader impacts of scientific
research are evaluated across a range of international research organizations.
This work was a continuation of a multi-year project of ours that had focused
on NSF's broader impacts criterion. In 1997 the NSF changed the criteria for
the peer review of research proposals: panelists would now review proposals on two criteria, intellectual merit and broader impact. 'Intellectual merit'
focuses on the potential for advancing knowledge; 'broader impact' evaluates
the likelihood that the research would benefit society. While serving on an
NSF panel in the early 2000s, one of us discovered that panelists were having trouble making sense of the broader impacts criterion – what the term
itself meant, and how it related to or even differed from intellectual merits.
We began digging into the topic, in a project that by 2008 we labelled the
Comparative Analysis of Peer Review (CAPR; Frodeman and Briggle 2012).

While not described as such, the demand for broader impacts at NSF
and elsewhere comprises an unprecedented philosophical and political

intervention into the evaluation of science. Under the auspices of making science more relevant the NSF has created an innovation that threatens the entire architecture of modernity. The unique status of science, its ability to end conversations by an appeal to the facts of the matter, turns on its perceived separation from questions of politics, ethics, and values. Bringing broader impacts into the evaluation threaten to undermine or even destroy this distinction (cf. Frodeman 2009; Holbrook 2009). Of course, fifty years of work in areas like the sociology of science, hermeneutics, and the philosophy of science had put the lie to simplistic claims of scientific objectivity as being separate from a wide variety of values. The net result of the broader impacts criterion was to bring this philosophic insight into the world of science policy.

In our second grant, the project team would survey the peer-reviewed and grey (governmental) literature in order to see how research teams and governments were evaluating the broader impacts of science. It would look at how metrics used in research assessment 'black-boxed' value judgements, and more generally explore the theoretical challenges facing efforts to identify broader impacts – for example, problems of the timeline and attribution of effects, as well as the question of negative impacts or 'grimpacts.' Altogether, the two projects formed a neat pair: one would think about the impact of philosophy, and another would conduct research into the philosophy of impact.

Socrates Tenured is in part the outgrowth of these (as well as earlier) NSF grants. This and the next two chapters report on our findings from the three domains of applied philosophy, environmental ethics, and bioethics. We've argued that philosophy is much less relevant than it should be in contemporary society. Now we want to see if these three fields form an exception, and whether or to what degree they provide any useful approaches to how philosophy can better contribute to society.

THE PROFESSIONALIZATION OF PHILOSOPHY

Applied philosophy should be understood as a response to the twentieth-century professionalization of philosophy. Socrates wasn't an applied philosopher, nor was Machiavelli or Descartes; the term only makes sense as a reaction to the institutionalization (that is to say, the disciplining) of philosophy within the modern research university.

Specialization, accreditation, professionalization … all of these, while the sine qua non of academic life, should be recognized as problematic to philosophic inquiry. The disciplines of the university are built upon claims of knowledge; but what does this imply about the status of philosophy, which once grounded itself in Socrates' profession of ignorance? The art Socrates practised was about questioning rather than providing answers, and challenging rather than colouring within the lines. He irritated and provoked, but supplied his conversation partners with few doctrines. He carefully distinguished being a friend of wisdom from actually claiming to be wise. And he certainly was no expert – with the exception of one (albeit crucial) topic, that of *eros*.

Socrates Tenured has no pretensions of offering an elaborated history of philosophy. Nor does it seek to provide a full account of the varied roles that philosophers have played in Western culture – a task that has been done often enough. Our focus is on a terrain that has been strangely ignored – the consequences of the twentieth-century disciplining of philosophy, and the possible post-disciplinary way forward. Nonetheless, a few signposts are worth noting, with a particular regard towards the question of professionalization.

Contemporary philosophy has forgotten that philosophers once rebelled against institutional enclosures. Thinkers like Descartes and Spinoza, Leibniz and Locke weren't housed in universities, but were members of other, looser institutional structures: literary and scientific societies outside the universities such as the Royal Society and the Académie Française, and more generally independent scholars who operated as members of the Republic of Letters. Locke, for example, had a love-hate relationship with Oxford, and did much of his thinking at Exeter House in London with Lord Ashley, or wandering in France and (in exile) in Holland, discoursing with a wide range of political and intellectual figures. Such histories should raise questions of what business philosophy has in being part of the normal bureaucratic structure of the university. We believe it does belong within the university, but not without a better appreciation of the resulting perplexities and absurdities, along with a commitment to promulgating a wider set of institutional and social roles. Thus courses in ancient and modern philosophy should stop treating Plato and Aristotle as *avant la lettre* academics, and take seriously the implications of Leibniz turning down an academic appointment at the University of Altdorf.

Does the professionalization of philosophy only mark a negative development in the history of philosophy? We don't believe so. For while we find it problematic, there are also virtues that have resulted from professionalization. Kant represents a crucial moment in this story. A passage at the beginning of the *Grounding for the Metaphysics of Morals* (1785) illustrates his anti-Socratism:

> All industries, crafts, and arts have gained by the division of labor, viz., one man does not do everything, but each confines himself to a certain kind of work that is distinguished from all other kinds by the treatment it requires, so that the work may be done with the highest perfection and the greatest ease. Where work is not so distinguished and divided, where everyone is a jack of all trades, there industry remains sunk in the greatest barbarism. Whether or

not pure philosophy in all its parts requires its own special man might well be in itself a subject worthy of consideration. Would not the whole of this learned industry be better off if those who are accustomed, as the public taste demands, to purvey a mixture of the empirical and the rational in all sorts of proportions unknown even to themselves and who style themselves independent thinkers, while giving the name of hair-splitters to those who apply themselves to the purely rational part, were to be given warning about pursuing simultaneously two jobs which are quite different in their technique, and each of which perhaps requires a special talent that when combined with the other talent produces nothing but bungling?

Kant expresses the attitude that has guided the field's intellectual labours since the late nineteenth century: knowledge, including philosophic knowledge, consists of a series of domains best left to specialists. If we parse Kant's claims we find two reasons for embracing specialization. First, it confers authoritative knowledge or expertise. This is in contrast to the philosophic amateur who is (like Socrates?) a hopeless bungler. Second, it enhances autonomy. To develop a clearly demarcated realm of knowledge is to establish a domain that can be self-governing, driven by its own imperatives, and judged by its own standards. The converse in this case, the downside of not professionalizing, would be heteronomy – having outsiders legislate your will. This is known more colloquially as 'selling out'.

Of course, it is possible to bungle one's thinking. But the point is a difficult one to hold onto, for expertise has an inescapably sociological component. Knock-down arguments are a rarity in philosophy; more often, each argument elicits a counter argument, ad infinitum. Moreover, standards of scholarly expertise contain an element of arbitrariness and social construction – and thus of heteronomy: after all, what counts as expertise in a given field is the product not (just) of a pure self-legislating will, but is also a matter of such external stuff as the amount of resources (money, time, personnel) that have been devoted to examining a particular topic.

This also highlights Kant's ontological assumption that labour can be divided along some pre-existing, objective categories that we can cut topics, and nature, at the joints. But the objects of study and standards of acceptable discourse (not to mention the very notion of what should count as a profession) shimmy and shake with history. 'Expertise' presupposes disciplinary boundaries – where experts are legitimated in ignoring related areas of research. But by what logic does the scholar of Aristotle decide whether the next increment of study is best spent rereading one or another part of Aristotle's corpus, rather than a commentary on Aristotle; reading a related author, Plato for example, or a secondary source on the same; turning to a cultural (or military, sexual, ecological, etc.) history of the era; or in (re)reading Thucydides, Aeschylus, Euripides, and Sophocles, or any number of secondary sources on these latter? Appeals to expertise offer no help here. The boundaries of one's study trail off, and one is left to exercise judgement. Indeed, scholars in the same sub-specialty choose different paths all the time, creating overlapping magisteria. And as they wander in different directions, mixing different approaches, they might start to look (even to one another!) like, well, a bunch of bunglers.

Kant's commitment to autonomy rests on firmer grounds. Independence is central to the function of the modern university: reason must be autonomous, demonstrating the self-coherence of an abstract system not subservient to outside influences or forces. Put differently: philosophy, or at least parts or moments of philosophy, should not be forced to live by a time that's not its own. To submit to such claims for timeliness is to subjugate oneself to outside forces, or as Kant called it, to be heteronomous.

This point represents a significant challenge to field philosophy. How does field philosophy – or any attempt philosophers make at being relevant – keep from slipping into becoming a hired gun for one or another interest group, or in other words, becoming sophistic? Our response is to move from Kant to Hegel and to treat the matter in a dialectical fashion, cycling from heteronomy to autonomy and back again. Field philosophy begins with a problem as defined by others (heteronomy) but then preserves its own independence in reacting to it by the dictates of reason (autonomy). Field philosophy submits itself to the needs for timeliness of its audience (heteronomy) but then reflects in leisure on the larger implications of the lessons learned (autonomy). Finally, field philosophers are seconded to other groups and organizations (heteronomy) but after a period of time return to the department to renew the sources of their insights. The problem with Kant's position, then, is that it is a partial truth that has become the entire truth. Disciplinary philosophy is essential – but must be supplemented by time spent in the field.

The Hegelian trope is one way to frame this point, but we can also make it differently by looking at the nature of professions. Michael Davis (2005) offers the following definition: "A profession is a number of individuals in the same occupation voluntarily organized to earn a living by openly serving a certain moral ideal in a morally permissible way beyond what law, market, and morality would otherwise require" (p. 443). Engineers serve the ideals of utility and safety, while doctors serve the ideal of health. But they don't just serve these ideals; they do so according to standards of education, skill, and ethical conduct. And these standards are their own design and are under their own control. Here again is the dialectic of serving others, but doing so with an integrity defined by autonomous norms. Similarly, an applied anthropologist, chemist, or lawyer may serve clients, but they do so in accord with professional standards. To fail at that constitutes malpractice or misconduct.

Kant would claim that the philosophical profession he envisions serves the ideal of truth. But as the wholly inward-looking logic of his argument took institutional shape in the form of the discipline, this ideal became hermetically sealed. It became 'truth-for-fellow-adepts' or 'contemplation-with-fellow-adepts.' Yet physicians don't just heal themselves. Engineers don't just make useful tools for other engineers. We see no reason why philosophers need to be trapped in a solipsistic model of professionalism. Indeed, this is a crucial distinction that T.S. Eliot made, between auto-nomos (self-legislating) and auto-telic (self-justifying). In the latter case a profession not only controls their own affairs but takes those affairs to be worth pursuing for their own sake (see Fuller 2002). Academic philosophers have too often conflated these terms, supposing that the only version of autonomy available to them was one that is autotelic: the only way to self-govern is to position ourselves in a world beyond utility. Field philosophy opens up this distinction, offering a model of philosophy that is justified at least in part by external considerations, while nonetheless governed at least in part by internal standards. You can philosophize with and for the people without slavishly (instrumentally, toolishly) saying only what they want to hear. You can be Socrates rather than a sophist.

Lamentably, this distinction between different elements of philosophy has been lost. Public philosophizing was replaced with Kant's version of freezer-wrapped erudition, which has deteriorated into a dogmatism of small consequences. In his 1784 essay 'What is Enlightenment?' Kant argues that those who hold a "civic post" can only make the "private use of reason." The soldier cannot question orders and the pastor cannot question church doctrines. Only acting as a "scholar," that is as a member of a "society of citizens" can someone make a "public use of reason" to question authority and engage openly on the important issues of the day. Only speaking "for himself" (not qua pastor or soldier) can he challenge the church or general. But Kant treats philosophers like pastors speaking with their disciplinary congregation rather than as scholars in broader society. There is no reason why one can't both hold a civic post (profession) and make public use of reason. That is, philosophers can speak to broader audiences – and not just "as a citizen" but as philosophers.

By the early twentieth-century philosophers had become the institutional sons and daughters of Kant. Nonetheless, a few early twentieth-century thinkers recognized these points and fought against the tide. We have already cited Dewey in this regard; Bertrand Russell is notable as well; and as Bordogna (2008) notes, William James sought to "alter the way that many American philosophers understood their discipline and challenge the professional self that many of them championed" (p. 14). James thought that philosophers in the early twentieth century were becoming too obsessed with disciplinary routine and too hungry for scientific respectability as they sought to avoid being seen as mere amateurs. In these ways James echoed Nietzsche, who had earlier spotted something deeply unphilosophical about nook-dwelling specialization.

A survey of websites of the oldest and most prestigious programmes of philosophy in the US reveals no awareness of the significance of the new institutional form of the department. The University of Chicago states that its Department of Philosophy was founded in 1894, with John Dewey serving as its first chairman, but only notes in passing that it is "one of the oldest in the United States." Princeton Philosophy describes its history as going "back to the College's earliest years, long before the present departmental structure was instituted in 1904" – without commenting on the reason for or significance of creating that structure. Michigan traces its origins to 1843, with the creation of a professorship of moral intellectual sciences, but does not mention the inauguration of the department itself. The same is true at the Cornell department website, which says only that the department was founded in 1891 with an endowment from Henry W. Sage. Programs will list their illustrious ancestors – Johns Hopkins mentions Peirce, Dewey, and Lovejoy – but these thinkers are free-floating generators of insights seemingly unaffected by their institutional housing. New systems were changing the workplace in the Age of Ford, and within the university the older ideal of *Bildung* was giving way to a mélange of disciplinary goals; but the possibility that the department represented a similar movement is passed over in silence. Ford's productionist metaphysic came to hold sway as one academic function came to dominate all the rest: produce new knowledge. Philosophers fell under its trance, embracing a new existence as professionals and experts. Donoghue (2008) notes that universities are characterized by

> a collective behavior that ironically duplicates the very corporate values from which we humanists wish to distance ourselves. Since the 1970s, we in the humanities have adapted to the conditions of our profession by developing a culture as steeped in the ethos of productivity and salesmanship as anyone might encounter in the business world. (p. 26)

While James fought a losing battle against this inward-looking professionalization, a more outward-looking posture held on for a while in some quarters. The Philosophy of Science Association (PSA, founded in 1933) adopted a broad, public-spirited mission statement in 1946 of "furthering the study and discussion of the subject of philosophy of science, broadly interpreted, and the encouragement of practical consequences which may flow therefrom of benefit to scientists and philosophers in particular and to men of good will in general." But just ten years later the PSA had turned inward, limiting their mission to "the furthering of studies and free discussion from diverse standpoints in the field of philosophy of science, and the publishing of a periodical devoted to such studies in this field" (see Douglas 2016).

McCumber (2001) puts another historical framing on these matters, arguing that the toxic political environment of the 1950s left a lasting impression on philosophers of the time. Faced with McCarthyite political threats, philosophers abandoned critical social commentary and instead adopted a "hidden protocol" to guide their work: the principle that "philosophy must, in order to be rigorous, be *restricted* to investigating the truth of sentences or propositions" (p. 163; emphasis in the original). This protocol became central to how academic philosophy was pursued. Philosophers have "never openly debated it" since its adoption, even though it continues to shape the content and character of contemporary philosophy (p. 163).

Dewey and James lamented the insularity of professional philosophy decades before McCumber's "hidden protocol" was born; but McCumber does add to the story by offering reasons for why philosophy became even more inwardly focused during the early years of the Cold War. Philosophy was already a specialized profession by the time McCarthy rose to power; absent this fact the hidden protocol would have got less institutional traction over subsequent generations of philosophers. To be a real philosopher meant passing muster with one's disciplinary peers. Thus when applied philosophy arose in the 1970s, its defining anxiety was to prove that it was sufficiently professional, as defined by the disciplinary standards of the research university.

McCarthy era politics, then, exacerbated existing professional tendencies towards specialization and internal standards for progress. Philosophers were already governed by institutional structures promoting original, peer-reviewed research aimed at other philosophers; building on this, the 1950s saw the rise of conceptual analysis, explicitly modelled after the disinterested, objective scientific observer in the pursuit of technical knowledge claims. And by some standards academic philosophy across this period was a success. A booming higher education industry during the Cold War meant colleges and universities were adding graduate programmes in philosophy, and tenure-track positions were abundant. But by the measure of public engagement, philosophy was a failure. Kuklick notes that "professional philosophy had become a way, not of confronting the problem of existence, but of avoiding it" (1977, p. 572). While internally successful, the discipline "could not head off accusations of narrowness and irrelevance" (Kuklick 2001, p. 261).

THE RISE OF APPLIED PHILOSOPHY

The 1960s were a time of rising professionalism within philosophy. At the same time, some philosophers started to buck the status quo. They were responding to larger cultural dynamics – social and political upheavals such as the Civil Rights Movement, the Vietnam War, and the backlash against environmental degradation. Philosophers found themselves facing repeated challenges from students about the value of philosophy for public life. Some of the ideological force for these calls was drawn from existentialist philosophers like Sartre and the social critiques of Marcuse and the Frankfurt School (Kuklick 2001). At least some challenges to the philosophical professoriate, then, were grounded in the writings of professional philosophers – although via figures like Ivan Illich, who remained marginal to the dominant trends within the discipline.

In 1970 Leslie Stevenson – best known for his work on emotivism – distinguished between pure philosophers concerned with "the scholarly pursuit of truth for its own intrinsic interest" and applied philosophers concerned with pursuing "answers which are needed" for determining what we ought to do (p. 259). Stevenson was critical of the view of philosophers as being only technical researchers, arguing instead that all philosophers should at some point do applied philosophy, the aim of which is to provide explicit guidance for action. By 1977 Jon Torgerson noted rising concern within the philosophic ranks that "philosophy ought to make contributions beyond its own specialized discipline. ... That is, philosophers have more to do than just their own technical analyses of other philosophers" (p. 215). Five years later, Louis Katzner criticized philosophers for having "eschewed the idea that philosophy is a tool for addressing the problems in living that confront individuals and society on a daily basis" (p. 32). For Katzner the scientistic stance towards pure research characteristic of analytic philosophy led philosophers to immerse themselves in purely theoretical problems like "action theory focused upon what counts as an action, not how to translate thought into action" (p. 32).[5]

A number of definitions of applied philosophy have been proposed since Stevenson's formulation. Jonathan Dolhenty's description (2008) is typical: applied philosophy "is the application of those principles and concepts derived from and based on philosophy to a study of our practical affairs and activities." Note the assumptions built into this view: theoretical philosophy first settles on the correct principles and concepts; then the principles are applied to a given situation. There is no hint that the question of how to 'apply' philosophy is itself a philosophic question, or that the flow of the argument could move in the opposite direction, from our embeddedness in a situation to the derivation of philosophic principles. So we find Onora O'Neill (2009) claiming that applied philosophy is "a genre of academic writing that seeks to identify and vindicate normative, action-guiding claims, and then relate them to the facts of situations or cases."[6] Similarly, David Archard (2009) states: "As the title suggests, [applied philosophy] implies the application of principles, already known and established as independently true, to particular domains."

One does find dissenters. These even include O'Neill, who in the same article makes the case for the importance of practical judgement and a critique of simplistic notions of applying principles: "If we think of applied ethics merely as discussing principles and the types of situations in which they might be applied, we say too little about the practical task of working out how principles are to be enacted in those situations, and how conflicts and potential conflicts between them are best handled or averted" (2009, p. 229). More pointedly, Fullinwider (1989) argues, "Philosophical theories, in contrast to philosophical training, are the wrong tools to bring to practical policy questions. Philosophical theories answer philosophical questions, not practical ones." But the debate over the structure of applied philosophy has remained quite primitive. There is a striking lack of meta-philosophical commentary on the nature, goals, and impacts of applied philosophy, as we will see below.

Applied philosophy was institutionalized in the 1980s with the founding of two flagship journals. The mission statements of both offer extra-disciplinary justifications for their existence, by noting the larger societal relevance of applied philosophy. In 1982, the *Journal of Applied Philosophy* (*JAP*) said that its aspiration was to provide "a unique forum for philosophical research which seeks to make a constructive contribution to problems of practical concern." In 1984, the *International Journal of Applied Philosophy* (*IJAP*) affirmed its commitment "to the view that philosophy should be brought to bear upon the practical issues of life." But one finds very little reflection about how these goals were to be achieved, whether such goals might affect the nature and standards of the philosophical enterprise, or whether academic journals are the most appropriate venues for achieving those goals. The announced goal of the applied philosophy movement was to be useful to society at large as well as to develop nuanced philosophical commentary. But there was little or no organized discussion of what the goals of the movement exactly entailed, how they might be achieved, or how progress could be measured.

Armed with our NSF funding and with the help of graduate student Kelli Barr, we set out to survey the applied philosophy literature looking for answers to these questions. Our survey focused on five journals: the two flagship journals in applied philosophy just mentioned, the *Journal of Applied Philosophy* (1982) and the *International Journal of Applied Philosophy* (1984); *Metaphilosophy* (1970); and two journals lying at the intersection of philosophy and policy, *Philosophy & Public Affairs* (1971) and *Philosophy & Public Policy Quarterly* (1987). Collectively, at the time of our study (2015), these journals had published 4,561 peer-reviewed articles on the philosophical elements of law, politics, science, economics, culture, environment, medicine, education, and policy.

For each of these articles we read the title, abstract, and (when available) the first page. An article was flagged for a full reading and analysis if either of the following subjects was broached:

- The meaning of 'applied philosophy' or 'applying philosophy'. This included boundary work on what applied philosophy was or was not, the methods of applied philosophy, and defences of applied philosophy as properly philosophical work.
- Descriptions of progress made, or failures noted, in integrating philosophy within specific contexts and for specific non-philosophic audiences. This included looking for accounts of lessons learned (i.e., 'best practices') in working with these audiences.

Our investigation, then, rested upon a distinction hardly found in the literature – between offering insights *about* a topic of interest to non-philosophers, versus making practical efforts towards actually *integrating* these insights with the ongoing work of non-philosophic audiences.

This is a point about audience. The former approach, whatever virtues it might have in terms of insights into practical problems, involves passing muster with a group of philosophical peers (most obviously, because the writing will appear in a peer-reviewed journal). In contrast, the latter point entails going out into the world: talking to people, attending meetings, getting into the details of bureaucratic opportunities and limitations, and, in general, becoming one voice among a number of competing perspectives. The latter task is often described as 'dumbing down.' And no doubt, it has its moments of boredom. But we have found that engaging non-philosophers on an ongoing basis raises a whole range of philosophic questions that philosophers have scarcely considered.

In our analysis we divided the 4,500-odd papers into three groups. Most common were articles that ignored the above distinctions and thus fell outside our two categories. We labelled these as 'disciplinary applied philosophy' – papers whose concern was with making philosophic points about a topic of interest, but without any discussion of how those points could actually be applied in specific contexts. This totalled some 98.6% of the papers. A paper might analyse the public debate or policy processes surrounding stem cell research – but never broach the question of how its analysis could play a live role in that debate. It might even develop recommendations – without any apparent attempt at seeing how those ideas could be taken up by decision makers.

The second group of papers, comprising fifty-five articles (~1%), discussed the first point above, the nature and purview of applied philosophy. Authors here reflected on what applied philosophy is and how best to accomplish it, its relation to other disciplines or to other areas of philosophy, or its relation to the whole of philosophy (see esp. Bowie 1982; Katzner 1982; Kasachkoff 1992). Such papers were especially common in early volumes of *JAP* and *IJAP*. And on one occasion, a special issue of *JAP* (in 2009) focused on the relation between philosophy and public policy. But while some authors in this special issue touch on the question of using philosophy in specific contexts, rather amazingly none did so in a sustained manner. In fact, in many cases, the primary concern was to defend the status of applied philosophy as compared with theoretical or 'pure' philosophy – arguing that applied philosophy should count as 'real' philosophy.

The third group of papers consisted of those that addressed point 2 above. These authors treat the application of philosophical insights as a philosophical problem in its own right and deserving of sustained reflection. This constituted a mere handful of papers – eight in total, out of more than 4,500. Let us be clear: of course it is possible that any number of the 4,500 papers have had an impact on the larger world; but *accounts* of this impact, reporting back to the discipline, were not making it into the pages of these journals.[7]

As an example in this last group, Kagan (1985) described her impact as a legislative director for a California congressman – a position she secured as a result of being an APA congressional fellow. Kagan's goal was to use her prior philosophical work on animal ethics to inform the content of animal protection legislation. As evidence of her impact, she described how a major piece of legislation she worked on was ultimately signed into law. In another article, Hare (1984) reported on his experience as a fellow, where he authored a number of pieces of legislation, some of which were eventually passed as law. He described this work as a natural extension of his philosophical research into contemporary social and political theory.

APA fellowships were offered from 1979 until 1984, supported by two three-year grants from the Andrew W. Mellon Foundation.[8] Bill Puka (1986), also a former fellow, recounts that the fellowships were modelled after the Science Policy Fellows programme at the American Association for the Advancement of Science. Participants were guaranteed funding, but were required to secure a position in the office of a Congress member through their own efforts.

Another noteworthy exception to the general trend: in its first three volumes (1985–1987), and then again in its ninth, the *IJAP* featured a type of article called 'Reports on Applying Philosophy'. In these reports philosophers reflected on their experiences as 'expats' in having left the land of philosophy. Some had found employment in computer programming, banks, hospitals, congressional offices, or other academic disciplines. Several reports argued that philosophers should bring their skills to bear in a particular industry or discipline – technical writing, for example, or psychometrics (Girill 1984; Colberg 1986). Others commented on the kinds of skills or attributes philosophers could bring to complex policy problems such as animal protection and welfare or ethics education for medical professionals (Kagan 1985; Fleetwood 1987; Snowden 1982). Of the fifteen reports published, five broached the question of what, if any, impact their presence had upon the hospitals and legislative settings where they worked.

It's likely that societal factors played some role in driving the degree of reflexivity among applied philosophers. For example, by the mid-1970s, the job market in philosophy had taken a significant turn for the worse. This prompted the formation of the APA's Subcommittee on Non-Academic Careers for Philosophers. In 1977, the subcommittee published a report in *Metaphilosophy* summarizing the deteriorating job market for philosophy PhDs. Citing a report from the National Board on Graduate Education, the subcommittee noted that 69% of philosophy PhDs awarded in 1974 landed teaching positions at a college or university, down from the more than 90% placement rates reported in the 1960s. Projections for future placement were grim: by the 1980s, placement could drop as low as 15–20%, according to the Carnegie Commission on Higher Education. Further, the Carnegie Commission warned that the "shortfall of academic openings would be felt most severely by those disciplines that have few attractive job alternatives outside of higher education" (APA 1977, p. 233).

In anticipation of philosophy PhDs turning to the non-academic job market, the APA report featured three case studies in non-academic employment for philosophers. Graduate students in philosophy, the report explained, "need not despair and expect that they are doomed to intellectually stifling, non-professional positions for which they may have been qualified upon graduation from high school. There are career alternatives to teaching that are both intellectually stimulating and professionally challenging to philosophers" (p. 234). Those with PhDs offered accounts of how their philosophical skills prepared them for policy research at consulting firms, for creating computing languages and programmes, and for designing policies for legislative mandates like the Equal Employment Opportunity Act. Both the APA report and the 'Reports on Applying Philosophy' made real efforts to grapple with questions of philosophy's broader relevance.

But no other journal in applied philosophy featured anything similar to the 'Reports on Applying Philosophy'. And these reports disappeared only a few years after the experiment got underway. Their (unexplained) discontinuation points towards the pull of disciplinary capture. Of course, continuing to publish 'Reports on Applying Philosophy' is no guarantee of impact, but it would have helped to embed the question of impact within the philosophical literature.

Finally, in an essay that appeared in 2011, Ben Hale argued that philosophers need to develop tools like those used in schools of policy studies. Too often, philosophers have something like a 'search image' that trains their vision to see only what they take to be the properly philosophical dimensions of a problem. They then extract that material and treat it in abstract theoretical ways, which leaves decision makers in the lurch when it comes to making decisions sensitive to the complex contexts involved. He concludes, in a way that echoes Thomas More in *Utopia*, that "philosophy is relevant indeed. It's just not relevant in the right way" (Hale 2011, p. 220).

Our survey revealed a remarkable absence in the applied philosophy literature. While a great deal of this work is in principle relevant to a wider set of audiences, we found few accounts of the impact that applied philosophy has had on those audiences. Applied philosophy journals are of course the appropriate place to find research *about* social problems. But they are also a natural venue for reports on success stories and best practices for *integrating* research with various stakeholders. But beyond the special set of reports in the early days of the *Journal of Applied Philosophy*, essentially none of these practices exist.

Summarizing our results, the mission statements of applied philosophical journals announce that this work had two goals: disciplinary and extra-disciplinary impact. Moreover, it is clear from reading these articles that applied philosophical research wants to be relevant to STEM and policy audiences. Yet, we found only a handful of discussions – 8 out of 4,561, or 0.18% – about the various types of impacts, accounts of success or failure in having a broader impact, best practices for how to have impacts, or metrics for evaluating impacts. We did find indications that the accounts of philosophic impact missing from applied philosophy journals might be found across a diverse range of non-philosophical journals (e.g., in *Conservation Biology* or the *Journal of Law, Medicine and Ethics*). But even here the recursive move seems absent – that is, there is little reflection on *how* to have impact.

In sum, philosophers are doing socially relevant research, and this research may even be having an impact. But philosophers rarely reflect on their work in print, or seek to demonstrate or measure their impacts to their philosophical colleagues. There are essentially no systematic accounts of their successes or failures or of what constitutes best practices for achieving broader impacts.[9]

Now, in presenting these accounts in different venues we have heard various criticisms. Some have claimed that we have been looking in the wrong place – accounts of impact would appear elsewhere, within stakeholder venues rather than in the pages of disciplinary academic journals. Fair enough;

but shouldn't those accounts make it back to the applied philosophy journals, so that philosophers can learn from one another? We also grant that some philosophers have been so successful in their extra-disciplinary work that their impacts on other fields may rival their influence within philosophy. Examples of this include Michael Walzer within law, Baird Callicott within conservation biology, Dan Hausman within economics, Daniel Dennett within cognitive science, Martha Nussbaum within development studies, and Paul Thompson within agriculture.

We are not claiming that there are no philosophers doing important work in applying philosophy. But it is clear that this is a decidedly a fringe element alongside a disciplinary orthodoxy that encompasses even the vast majority of applied philosophers. In particular, we are claiming that there is a massive failure of reflexivity about *how* to do socially engaged research – on how to work *with* stakeholders on an issue and not just talk *about* that issue with philosophers. Applied philosophy journals should be the places where we see philosophers reflecting on such work, identifying best practices and lessons learned; yet we found essentially none of this. As a result, efforts to institutionalize alternative philosophical practices remain undeveloped. So too are efforts to train the next generation so that engaged (or what we call 'field') philosophy can become a respectable part of twenty-first-century philosophy.

All of these points are absent from the literature. This entire topic requires a taxonomy of impact, separating out several different types of phenomena that need to be examined:

- Definitions of 'impact' (effects, influences, nudges, etc.)
- *Intent to* have impact, which could be assumed or planned
- *Actual* impacts, which can be accidental or deliberate, and more or less direct
- *Proof* of impact, which consists of stories or numbers (themselves based on stories)
- Accounts of *how* to have impacts

What we were looking for in our survey was primarily in the last category: accounts of how to have an impact beyond a disciplinary cohort. We maintain that applied philosophy journals are a reasonable place to expect to find these accounts. Of course, all these categories co-determine one another. For example, in order to intend to have an impact or offer an account of how to have an impact, one needs to know what counts as an impact and how we'd know if the impact actually happened. And to claim that one has had an impact implies that one has successfully fulfilled an intention and could offer an account of how that impact came to happen.

DISCIPLINARY CAPTURE

In rocket science, escape velocity is the speed needed to break free from the gravitational attraction of a massive body without further propulsion. Achieving it takes lots of energy, especially if the gravitational pull is strong. That's why most things on the planet stay on the planet. This is a useful concept for thinking about applied philosophers. They have wanted to escape the ivory tower and have a real-world impact. They've been trying to break free of forces that keep them grounded within isolated and insular discussions. But this mission has been a failure. Yes, work from the 1980s has produced a great deal of subtle philosophic analysis of practical problems. There *have* been effects, but they have mainly resulted from the slow molecular process of passive dissemination. Outside some notable exceptions, the field has failed in terms of having direct extra-disciplinary impacts.

This is because philosophers have not been aware of the gravitational pull of the massive body against which they must struggle. Philosophers thought they were trying to break free from the chains of an abstract discourse (i.e. 'pure' philosophy), and that talking about real-world issues would provide enough fuel to achieve escape velocity. But the gravitational pull holding them back was institutional rather than discursive in nature. Hale has a point about the lack of policy tools within philosophy; but the problem hasn't been primarily with the content of philosophical thought or its lack of a toolish nature. It wasn't just about *what* has been said, but also about *who* they have been speaking to.

The institution holding them back was the discipline-based university. Disciplines do a great job of developing new knowledge at high professional standards. But they do a poor job at transmitting that knowledge to society. Consider again the trope 'applied'. It indicates that the philosopher first does the intellectual work in specialized journals for one's disciplinary peers. Afterward, that work is supposedly 'applied' to society as a finished product. The passive voice is intentional here, because there is no account of who does the applying or how or in what ways success would be defined or measured. In the few self-conscious accounts we did find, the 'applied' metaphor is all wrong. Philosophers involved in the midst of a policy issue are not applying paint to a house. They are interjecting, adjusting, reacting, tweaking, listening – they are doing philosophy on the fly, *in media res*, in the interstices of thought and action.

Although it may have challenged 'pure' philosophy in some ways, applied philosophy left its institutional core intact. Applied philosophy didn't break from the disciplinary research model – a model premised on the separation of knowledge production and use. Indeed, it merely appropriated this model as part of its self-understanding about what makes philosophy 'applied'.

One first does the philosophy and then, somehow, it gets used by others. This demarcation between philosophy and society turns on the same underlying logic, whether the work is 'pure' or 'applied'. The applied turn was a turn *within* the disciplinary model.

The applied model of philosophical scholarship has the same faith in the passive diffusion model of knowledge transfer: peer-reviewed articles somehow lead to societal benefits. Insights diffuse like a concentrated gas. Or perhaps like how the economic benefits achieved by the rich are supposed to trickle down to the rest of society. But it's just this kind of nonchalant hand-waving that precipitated the accountability culture now taking hold of the academy. Philosophers feel like they have concluded their intellectual tasks and professional obligations when they have added another peer-reviewed publication to the reservoir of knowledge. Whether and how that knowledge gets used – well, who can say? But such faith-based impact stories are no longer adequate.

In the introduction to the 2009 special issue of the *Journal of Applied Philosophy* Archard and Mendus speak of the recent expansion of applied philosophy:

> In 2009 the philosophical lens has widened somewhat and all our contributors are concerned not simply with the application of philosophy to practical problems (though they are of course concerned with that), but also with the prior question of what it *means* to apply philosophy to practical problems. (their italics)

Excellent! But then the discussion within the special issue never returns to the question. Yes, authors skirt the edges of the problem – for example, Buchanan raises questions about what he calls the "Commission Paradigm" of applied philosophy, and Archard speaks of his "attachment to a version of what is known as 'bottom up' applied philosophy, one that starts from the facts of the real world." But this turns out to be a very abstract form of the "real world." The articles lack any account of anyone's actual engagement in the particular problems of people in real time – as one of us offered in a recent book (Briggle 2015). Nor is there any reflection on how or to what degree the profession's standards of rigour would need to adapt to practical exigencies, or how actual engagement would affect academic standards for tenure and promotion. Instead of talking about abstract notions of free will in peer-reviewed journals, applied philosophers talk about concrete problems of, say, euthanasia or endangered species. But they still talk in the pages of peer-reviewed journals, and without including an account of how these insights are supposed to be taken up by people outside the academy. Absent is any reflection about how to actually get involved with the stakeholders in particular policy processes, how to effectively interject insights into conversations, or how to track the impacts of one's efforts.

Philosophers and humanists generally have suffered from disciplinary capture.[10] The first step to breaking out of this gravitational pull is to name it. From there, philosophers need to change practices and incentives so that they write for and work with a wider set of peers; in particular, they need to cultivate a demand-side aspect to their work. Moreover, those few who have travelled to the great beyond need to occasionally visit their home planet to train the next generation. For as a community, we cannot sustain escape velocity if our rare successes remain one-off ventures rather than a self-conscious, collective creation of a new paradigm for philosophy.

NOTES

1. "Now if you take away from a living being action, and still more production, what is left but contemplation? Therefore the activity of God, which surpasses all others in blessedness, must be contemplative; and of human activities, therefore, that which is most akin to this must be most of the nature of happiness." *Nichomachean Ethics*, Book X, Chapter 8.

2. Or not quite rejected: the National Science Foundation Act left the door open for the funding of "other sciences." The attitude of many (natural) scientists at the time towards the social sciences is summed up by a National Science Board member: "We have to face up to the fact that the social sciences – except for a few extremely limited areas – are a source of trouble beyond anything released by Pandora." From 'The National Science Foundation: A Brief History', https://www.nsf.gov/about/history/nsf50/nsf8816.jsp.

3. It should be noted, though, that Vannevar Bush, who authored the conceptual blueprint for the NSF thought that the humanities were vital. The National Endowment for the Arts and the National Endowment for the Humanities were both founded in 1965 – perhaps the high-water mark of American liberalism.

4. More on this below. We continue to submit applications for funding this research.

5. Continental philosophy was not appreciably more successful at doing practical philosophy. But as the dominant type of philosophy in twentieth-century America, analytic philosophy was the most commonly criticized.

6. O'Neill goes so far as to claim that "writing in applied ethics has to abstract from the details of actual cases, in favour of discussing schematically presented *types* of situation or case" – a view we will stand on its head.

7. Here is another way to approach these issues: begin with applied philosophers who are known to have had broader impact; see where they have published; and then survey those journals for philosophic impact.

8. See http://mellon.org/grants/grants-database/grants/american-philosophical-association-inc/37900007/.

9. Our literature review supports this claim, but our anecdotal experiences do as well. For example, Paul Thompson of Michigan State, who has long done socially relevant philosophy, said recently that he has never taken the time to write an account of how he performs his work – he has been too busy doing the work to reflect on it.

10. A word about the term: a colleague's (somewhat differing) use of the term dates back to 2009. See Brister, Evelyn. 'Disciplinary Capture and Epistemological Obstacles to Interdisciplinary Research: Lessons from Central African Conservation Disputes', *Studies in History and Philosophy of Biological and Biomedical Sciences* (2016).

Chapter 4

Environmental Ethics

Imagine a philosopher, an environmental ethicist, who wants to make a difference in the world. He comes upon a topic, a wind farm proposed for Nantucket Sound. He finds that the debate has been framed in terms of economics and science, even though he thinks the issue seems equally about aesthetics and metaphysics. Some see wind turbines spread across Nantucket Sound as despoiling a beautiful seascape; others view the huge metal structures as violating the natural character of the Sound. On the other side, some think the wind turbines are a magnificent symbol of the end of fossil fuels, while others see the Sound, like all nature, as mere stuff to be used however we wish. Our philosopher argues that even though these philosophical attitudes are largely driving the debate, they have remained hidden because of a set of philosophical prejudices: for beauty is only a matter of subjective taste, and metaphysics – well, what is metaphysics, anyway? And so numbers have set the terms of the debate – the cost of the turbines, the cost of electricity per kilowatt-hour. These numbers, however, have been functioning as stalking horses for other fundamentally philosophic concerns.

Our philosopher writes up this account, and after some time and effort the essay is published in the peer-reviewed pages of an academic journal (Briggle 2005). Ten years later, no one involved in the controversy has ever heard of him, and his paper has been cited once – by himself, in a later paper.

From a standpoint of social effects what is 'applied' about this work? Nothing distinguishes it from efforts in the core areas of philosophy. The paper resides in the same knowledge space, having been written for other PhDs in philosophy (the audience of the journal). It embodies the same institutional model for what counts as quality work, a peer-reviewed publication – never mind that it has no discernible impact on policy or indeed on any aspect of public discussion. Knowledge production has been divorced from consumption, as indicated by the fact that the essay included no scheme for how its ideas could actually be taken up by participants in the debate. The essay simply constituted another unit deposited into the reservoir of knowledge. Certainly, applied and 'traditional' philosophers write *about* different things; but they write in the same way, *to* the same type of audience. And in both cases the work takes place in splendid isolation – the solitary individual sitting before a computer, rather than in the field interacting with the actors in the drama.

Some will claim that there is nothing wrong here. And certainly this model can be defended on a number of grounds: the scholar needs to be free from the hurly-burly of everyday life; insights take time to disseminate; the author needed to write many more articles (and books!) on the topic; the lack of citations isn't a good indication of whether the paper had been read; or that this paper's failure is not representative of the field of environmental ethics.

But perhaps none of this is the real issue. For some will say that there is no 'model' operating here at all. What we're calling a 'model' is just what philosophy is – at least any philosophy worthy of the name. Philosophy consists of nuanced thinking between professionals, who necessarily live at a remove from the superficialities and incessant busy-ness of the masses. Public attention and larger effects are nice, but one does not judge philosophy in terms of how non-professionals respond to philosophic work.

Who gets to evaluate what counts as competent work in philosophy? Institutionally, for the last 125 years, the answer has been clear. PhDs in philosophy judge other PhDs in philosophy, particularly those who specialize in the area being discussed – Aristotle scholars for work in Aristotle, philosophers of science for work in the philosophy of science. But what about work, in any of these areas, concerned with issues that are relevant to the wider world? Shouldn't the people who are concerned with these issues have a voice in the matter? For how can philosophic specialists be the sole (or even the best) judges of efforts to relate philosophic insights to challenges faced by the NSF or a community struggling to make decisions about a wind farm?

Philosophers will continue to write for one another instead of the larger public until the discipline recognizes a plurality of standards for quality work. Our academic system simply isn't set up to judge whether philosophers are doing a good job at drawing out the philosophic dimensions of issues for non-philosophers. The system is only able to assess philosophic work by professional (more accurately, sub-disciplinary) standards. But this is to ignore a fundamental point of rhetoric – that the same claim can be true, insightful, or helpful from one perspective (or to one audience), and decidedly less so from another. Of course, the 'public' is riven by deep disagreements; there may be no shared standard to appeal to. Some will hate the philosopher's contributions; others might like them. But is this really a problem? Or is this simply the state of things in a pluralist universe, a fact that disciplines have allowed academics to ignore? For that matter, don't peer reviewers – even in the same sub-specialty – disagree all the time? Does the profession really have its own shared standard of excellence?

Philosophy in the twentieth century – a period that, in terms of institutional arrangements, we are still inhabiting well into the twenty-first century – assumed that it was another regional ontology consisting of experts trained and certified in the field. Note how the two concepts, expertise and regional ontology, are yoked: expertise is tied to the bracketing off of a limited domain. For one cannot be an expert in everything. But what if philosophy isn't (or isn't solely) a regional ontology? What if it is interstitial in nature, liable to pop up anywhere? Then we are left with the Socratic alternative: the philosopher's role is to profess ignorance, or at least scepticism, in terms of substantive beliefs, while emphasizing the importance of questioning and revisiting core assumptions. And while he was not much of a field philosopher, Heidegger emphasized these same points: for no one was more insistent about the interrogative character of thinking. Heidegger's difficulties with the profession on this score were severe enough for him to give up on its core identifying term (philosopher) and instead become a *Denker*.

These points are liable for misunderstanding, so let us be clear. Of course it is possible to be an expert in one or another domain of philosophy. There is no denying the mastery of someone who has devoted decades to thinking through logic or Leibniz. Clearly it's possible to treat philosophy in a disciplinary fashion, mastering a specific corner of the field. One can then make progress in the topic via recondite conversations with other specialists in the field. But what if one wants to bring these insights to a problem within the larger world? Now things get sticky. One still knows what one knows – ethics, or epistemology, say – but is it really as simple a matter as 'applying' a concept to a specific case? Or might one need to know details about that situation? If so, this will require you to leave your office and go meet with your interlocutors, learning enough about the state of affairs in order to make informed suggestions. But how many details? How many meetings? Do you need formal training in the area? Or perhaps only interactional rather than contributory expertise (Collins and Evans 2002)? How do we tell what constitutes sufficient knowledge of a case, and how do we integrate or blend different types of knowledge, the philosophical with the economic and the chemical? We have now opened up a whole new set of questions. Or: we can retreat to the assumptions of applied philosophy, where there are no real (philosophic) issues to be found in the application of philosophic insights.

These are difficult questions. In fact they lead to a topic that philosophers have given scant attention to, what might be called the philosophy of interdisciplinarity (Frodeman 2013; Frodeman 2017). But let us note that efforts to speak to different audiences in differing venues can result in something other than hackery. Rather, such efforts involve a different kind of 'hard': the difficulty of taking rhetoric seriously, balancing rhetoric and philosophy, insinuating philosophic thinking in an adept and timely manner. Brian Leiter (2007) unintentionally highlights our point when he argues that there are not different schools of philosophy, but only good and bad philosophy: "which 'camp' of philosophy could possibly be *committed* to less careful analysis, less thorough argumentation?" Our answer: a camp that is interested in being timely, within budget, and understood.

To sum up: applied philosophy was a twentieth-century movement that arose in response to disciplinary philosophy's lack of practical application, but did not realize what was entailed in that failure. One could claim that Thomas Aquinas (who wrote on marriage) and John Locke (who wrote on education and toleration) were applied philosophers – a list that can be expanded to include most philosophers across the history of philosophy. But this misses the point: these are pre-disciplinary examples, and applied philosophy was a response to the failures of disciplinary culture. The question facing applied philosophers was how to respond to societal demands for relevance. But it never occurred to them that answering this challenge should raise questions about their institutional housing and the *kind* of hard work that needed to be done. Doing applied philosophy, they thought, meant that they only had to talk about different things. Instead, they needed to talk in different ways to different people in different locations – to change the medium, not just the message. And they needed to do more than just talk. They needed to make philosophy into a practice.

The notion of practice does not come easily to Western philosophy. In the West, philosophy has consisted of Freud's 'talking cure' rather than being something that one enacts. But once philosophy is treated as a practice, as something one does as well as says, new questions appear. Most immediately, the institutional housing of philosophers becomes an issue. If philosophy exclusively consists of words, then we can operate as a brain in a vat. But if philosophy is a practice, location matters. And the deformational nature of being housed in a department becomes apparent. If we want to be effective we are going to have to leave the department.

Applied philosophy breaks into a number of areas – professional ethics (e.g. nursing), computer ethics, engineering ethics, bioethics, business ethics, environmental ethics/philosophy, and more. The previous chapter found that research appearing in applied philosophy journals was blind to, and thus fell prey to disciplinary capture. Do any of these subfields do a better job at having an impact, or reflecting on impact? Have they realized the importance of institutional housing? And if not, has disciplinary capture been the inevitable result? While a survey of all these fields is beyond the scope of this book, in this and the next chapter we explore two of these fields, environmental ethics and bioethics. They offer distinctive glimpses into the possibilities for avoiding disciplinary capture.

ENVIRONMENTAL ETHICS: AN INSTITUTIONAL HISTORY

Environmental philosophy is as old as Western philosophy itself. But 'environmental ethics' is a creation of the 1970s, and the product of disciplinary culture.

It wasn't called environmental philosophy across the millennia; different terms were used, like cosmology and *Naturphilosophie*. Nevertheless, a common thread connected Thales and Schelling. For thousands of years, philosophic speculation in the West was conducted within the overarching framework of nature. Philosophers sought to guide human conduct by bringing it into alignment with the patterns of the cosmos. The natural order was simultaneously a moral order, and ethics consisted in our getting in proper relation with universal, natural rhythms. Framed theologically, the order of the natural world testified to a benevolent providence that placed the Earth at the centre of its concerns. 'Environmental philosophy', then, lived out and about in the world as a politics and a theology as much as a personal code of conduct and theoretical endeavour.

The death of natural philosophy dates from the time of Darwin. *The Origin of Species* (1859) shattered the connection between the natural and the moral orders. Now, pace Paley, rather than being testimony to the artfulness of the Creator, the existence of the human eye reflected nothing more than a stupendous number of random events subjected to selective pressures across geologic spans of time. Stripped of theological connections and portents of larger meaning, natural science turned a cold eye towards the natural world.

Such was the state of things for one hundred years. When environmental philosophy was reborn in the 1970s, the field would be known as environmental ethics. This marked the fact that, as John Passmore wrote (1974), man had a responsibility *for* but not *to* nature: the metaphysics of nature had been replaced by physics, just as epistemology was naturalized and theology discarded (except as social science) by academic culture. But even here environmental ethics developed as a reaction to larger societal forces. Cultural moments like *Silent Spring* (1962) and the Santa Barbara oil spill (1969) led to Earth Day, the passing of the Clean Air and Clean Water Acts, and the creation of the Environmental Protection Agency – by Richard Nixon, no less. All of these events predate the inauguration of the academic field, which can be notionally dated by the first known course in environmental ethics, by Baird Callicott, in 1971. (Callicott reports that this course was itself prompted by Earth Day events on his campus, the University of Wisconsin Stevens Point.)

The first philosophy journal in the field, *Environmental Ethics*, was not founded until 1979 (four more have been created since). A new philosophical literature flowed across the pages of the journal. The 1980s saw the development of a rich ecosystem of thought. Important concepts were developed: intrinsic value and moral considerability, anthropocentrism, ecocentrism, and biocentrism, vegetarianism, and animal liberation vied with one another. An intellectual community was created. Clarity increased on a set of questions that had hardly been considered before – on species extinction, environmental restoration, and climate change.

The field had its successes. It has challenged long-held ethical norms. The environment had been the unthought of background of our activities; now, society asks questions about who – or what – deserves moral considerability. The moral standing of other species and even entire ecosystems has become an open question, challenging our Darwinian assumptions about animals and of nature itself. The fact that these discussions penetrated public consciousness reflects the successes of the trickle-down model. PETA (People for the Ethical Treatment of Animals) activists and advocates of personhood for great apes also showed the influence of the field – a far different result from the meta-ethical debates about free will or cognitivism versus non-cognitivism that occupied twentieth-century meta-ethics.

Yet, while the topics of environmental ethics were innovative and even paradigm breaking, the way that environmental ethicists went about their work stayed within disciplinary bounds. In the main, their work consisted of conceptual analysis rather than a kind of embedded practice. Environmental ethicists did not ask whether the institutional and disciplinary nature of philosophy affected the theoretical focus and practical efficacy of their work. And in remaining largely a disciplinary endeavour, environmental ethics consigned itself to the margins of both philosophy and society. Surely this made sense for a period of time; universities are an incubator of ideas, where new thoughts can be nurtured and worked out. But by the early 1990s even some environmental philosophers were chaffing at the insularity of the conversation. Bryan Norton, one of the founders of the field, argued in *Toward Unity among Environmentalists* (1994) that internecine theoretical battles were causing people to miss the chance to influence policy. Andrew Light and Eric Katz noted in their anthology *Environmental Pragmatism* (1996) that "the intramural debates of environmental philosophers, although interesting, provocative and complex, seem to have no real impact on the deliberations of environmental scientists, activists and policy-makers."

Others in the field voiced their concerns. Eugene Hargrove, the founding editor of *Environmental Ethics*, lamented the field's lack of a larger social effect. In a 1998 editorial 'Environmental Ethics at 20', Hargrove noted that, contrary to initial hopes, environmental ethics had remained an inward-looking group of thinkers engaged in theoretical debates (Hargrove 1998). In 2003 he raised the issue again. In 'What's Wrong? Who's to Blame?' he offered an account of why environmental ethicists have not played a larger role in policymaking (Hargrove 2003). While the perspectives of environmental ethics had been integrated in environmental science courses, and contributed to the creation of the new field of conservation biology, Hargrove found little penetration within the policy curriculum. On Hargrove's account, environmental ethics had been marginalized by the philosophic presumptions underlying policy analysis, whose understanding of human motivation had been taken from economics. Classical economics viewed humans as rational calculators focused on economic interests, and assumed that wants and desires only expressed subjective preferences. Classical economics also viewed the natural world as a set of resources to be transformed and upgraded into capital. Consideration of ethics was set to one side.

But Hargrove's account made it hard to discern cause from effect. Are environmental ethicists marginal players because of a philosophic prejudice? Or – a possibility not discussed by Hargrove – have bad philosophic assumptions taken hold because environmental ethicists have devoted themselves to high theory rather than working with a wide range of stakeholders over time? Perhaps if they had embraced a different kind of 'hard' and gone out into the policy fray, they would have had more success. But that, of course, would only have reinforced concerns of whether they should count as 'real' philosophers. Another factor Hargrove doesn't highlight is the absence of boundary organizations that could help to integrate the work of environmental ethics into public policy. Looking ahead to bioethics in the next chapter, this is one reason this field has been relatively successful: policymakers established institutional spaces where bioethicists can have a seat at the table where real-time, real-world decisions are being made.

What is undeniable is that by 1990, environmental ethics had become another academic specialty. Environmental ethicists inhabited philosophy departments, writing for a professional philosophical audience, publishing in specialized journals, and attending conferences populated nearly entirely by philosophers. They employed jargon and theories that assumed a shared expertise. Graduate students interested in environmental ethics were trained to work in philosophy departments rather than with scientists, activists, or policymakers. One found neither a practical (e.g. the development of internship programmes at the National Park Service) nor theoretical (e.g. how does the nature of our arguments change when our audience consists of non-philosophers?) questioning of the disciplinary model of knowledge production.

There were some notable exceptions. Andrew Light, for instance, after a couple stints as a professor, worked for a think tank and then the U.S. State Department; he still retains a foot in both the political and academic worlds. Michigan State's Paul Thompson has served on the National Research Council, worked with Monsanto, and advised the US government on issues surrounding agricultural ethics, all while teaching in a school of natural resources. And Kristen Shrader-Frechette has worked with scientists, engineers, policymakers, and NGOs on risk assessment, public health, and environmental justice issues. We could list other examples, and we ourselves have worked with the US Geological Survey and citizen's groups on questions surrounding acid mine drainage and fracking (among others). But the centre of the field remained resolutely (if unselfconsciously) disciplinary in nature.

A LOOK AT THE ENVIRONMENTAL ETHICS LITERATURE

We've noted that environmental ethics has had some notable success, via the trickle-down model, in terms of having wider effects; but what of attempts at taking a more active approach towards the world? There are a number of indices that could offer evidence of such efforts. One could interview stakeholders connected to environmental issues to see if they've been affected by work in environmental ethics. Or look for citations across a wide range of journals. Or search for mentions in the popular press. But as with the previous chapter, the approach we took consisted of a survey of environmental ethics journals, looking for accounts where research had made a difference to the larger world.

Our literature survey focused on five journals: *Environmental Ethics* (established 1979), *Environmental Values* (1992), *Ethics & the Environment* (1996), *Environmental Philosophy* (2004), and *Ethics, Policy, and Environment* (2011). These totalled 3,355 articles. Our survey (led, as before, by our graduate student Kelli Barr) employed the same method as the earlier review of the applied philosophy literature: reading the title, abstract, and (where available) first complete page of each article. Articles were read through to see if they featured either of the following points:

1. Boundary work on the scope and aims of environmental ethics, including discussions of its relation to the discipline of philosophy as a whole, to other disciplines, or to the public or private sectors.
2. Accounts of integrating the insights of environmental ethics within specific situations outside the academy, including discussion of the difficulties in doing so and best practices for successful interactions.

Once again the vast majority of the articles fell outside of these two categories. Some 98% of the articles limited themselves to conceptual analyses of issues in environmental ethics and philosophy: the nature of environmental values (e.g. intrinsic vs. anthropocentric), accounts or critiques of different worldviews, or the philosophical dimensions of environmental problems like endangered species or the Love Canal disaster. It was clear that these articles were written with a philosophical audience in mind, with no hint of anything like an implementation plan attached to their analyses.

About 1% of the total (thirty-three essays) fell into the first of our categories – boundary work seeking to identify the scope and aims of environmental ethics/philosophy. These essays raised questions about what environmental philosophers should be trying to accomplish, by what means they should do so, and the type of problems the field should address. For instance, in a 2007 article, Stephen Gardiner recounts the dissatisfaction many of his students and colleagues feel about contemporary environmental ethics, that it does not offer sufficient guidance on what to do now. Gardiner calls for an "ethics of the transition" to a more sustainable future. But he offers no account of what this should entail (Gardiner 2007).

The second category of articles, totalling close to another 1% (twenty-five, or 0.75%), raised questions of whether the theoretical orientation of environmental philosophy has borne practical fruit beyond the academy. For instance, in a 1979 *Environmental Ethics* editorial, Hargrove outlined his understanding of the purpose of the field: to develop graduate-level training in environmental ethics to influence those who would later become environmental professionals or policymakers. In 1993 Katz and Oechsli argued that a non-anthropocentric framework helps to resolve the apparent dilemma of local economic development versus environmental preservation. And in 2014 Kelbessa discussed how attention to the environmental values people already hold could inform and enrich public policy initiatives across several countries in Africa. But in all of these cases there are no indications that the authors are in fact involved in policy processes, or that there is any particular obligation for environmental philosophers to track or work out the results of their analyses. For example, if Katz and Oechsli's analysis resolves a pervasive policy problem, do they have a responsibility to test those findings in practice by working with policymakers? And should Hargrove's journal be tracking the influence of environmental ethics graduates on environmental professionals and policymakers – and then, perhaps, offering ways to strengthen that influence?

Many of the articles in this second category were authored by leading figures in the field. We have already noted the comments by Norton and Light, and seven of these essays consisted of editorials by Hargrove lamenting the fact that the field has had little practical impact since the founding of *Environmental Ethics* (see especially 1989, 2000, and 2012). We are the authors of several of these pieces (e.g. Frodeman 2006, 2007, and 2008). In 2007, Frodeman, along with Dale Jamieson, organized a workshop on the future of environmental ethics that raised the question of impact. Later that year a special issue of *Ethics & the Environment* published a series of commentaries from that workshop (with commentaries by Callicott, Norton, Hargrove, Gardiner, Ben Minteer, Holmes Rolston III, and others).

It is common for environmental philosophers to claim that they should be more involved with policymakers or other stakeholders. At the same time, they see problems with the relationship. Michael Bruner and Max Oelschlager (1994) put it this way:

> Since ecophilosophical discourse generally flies in the face of the prevailing social paradigm, and offers its ethical insights and ecological panaceas in a language that is not accessible to lay publics, it appears to be null and void from the beginning. In other words, environmental ethics appears to be incapable of moving a democratic majority to support policies leading toward sustainability. (p. 384)

Christopher Manes (1988) argues that environmental philosophy should not even be seen as philosophy. On the contrary, since environmental ethics "exists insofar as it acts, [and] not in the achievement of philosophic stability," it is better characterized as a task – "the fulfilment of which may actually and radically alter the systems of technical power that underlie environmental degradation" (Manes 1988, pp. 81–82).

Yet, these discussions stop short of seeking ways to evaluate whether or not the policy connection is being made. While a number of authors (e.g. Norton 2007; Minteer 2007) discuss in some detail the relevance of environmental philosophy to policymaking, our survey found no indications that the authors are actually involved in policy processes, or that this is an activity appropriate for environmental philosophers qua philosophers. Now, we know from personal interactions that (for instance) Norton and Minteer *have* published on philosophical topics in non-philosophical journals (e.g. the *Journal of Applied Ecology* and the *U.C. Davis Law Review*), and *have* worked with groups outside the philosophic community. But even as they (and others) reach out into the public, they do not return to the environmental ethics community to share accounts of their successes and failures. Until these efforts make it into the literature this work will remain marginal, lacking recognition and support within the field, where people can build off of one another's experiences.

Environmental philosophers may have focused more than applied philosophers on the question of how to apply their philosophical insights. In terms of raw numbers, the proportion of articles addressing this topic is slightly higher in the environmental philosophy literature than in applied philosophy.

Nonetheless, the overall message remains discouraging. Across the twenty-five articles in our second grouping, there is nearly unanimous agreement that environmental philosophy suffers from practical irrelevance. On their own account, environmental philosophers have failed to gain the influence that could bring about the social changes that they see as being necessary for addressing our environmental problems. But no one seems to have thought of treating this failure as itself a problem deserving of philosophic research.

Finally, we note that our explanation of this lack of practical effect – disciplinary capture – is the opposite of the criticism offered by some others, who have seen the field as suffering from a lack of philosophical rigour (e.g., Thompson 1990). Thus defences of applied philosophy that see it as being theoretically rigorous (e.g. Young 2004) miss the point. Or rather they exacerbate the problem: making applied ethics more 'rigorous' (understood traditionally, rather than the different kinds of rigour we advocate) will lead to more disciplinary insularity. As a consequence, the main result of environmental ethics – against its own better intentions – has not been to help those struggling with actual problems. It has been to refine the theories and methods of an academic discipline.

The fetishizing of rigour is part and parcel of the social and institutional trappings of a disciplinary model of knowledge production. While environmental ethics speaks to the kinds of 'external' problems now quite 'hot' across society (e.g. sustainability and resource consumption), it only speaks to them in academic terms. What a public policymaker, engineer, activist, physician, or economist makes of the work of environmental ethics is not the field's primary consideration – indeed, if it is considered at all. The environmental ethicist is satisfied when disciplinary peers approve of her Rawlsian framing of environmental justice, even if her work has no chance of landing on the desk of someone who is actually making decisions, say, about the siting of waste management facilities. Environmental ethics and theoretical philosophy share the same accountability gap, because they share the same disciplinary culture.

OTHER INDICES, AND ENVIRONMENTAL PHILOSOPHY IN THE FIELD

Our survey offers one view into how fields like environmental ethics are (or are not) thinking about questions of implementation and broader impact. This survey, of course, only constitutes an initial sweep; other indices need to be developed to get a better handle on the question of impact. There are signs that the philosophic community is becoming more attuned to these issues. For instance, Plaisance and McLevey have launched a project to map the use of the philosophy of science within the natural and social sciences from 1952 to 2014.[1] And we, in league with Britt Holbrook, have been working on questions of broader impact (both within philosophy and across the academy since 2005 [e.g. Frodeman 2006; Holbrook 2009]).

In 2015, we experimented with another way of taking account of how, or whether, philosophy was thinking about questions of implementation and impact, by attending philosophy conferences that share an interest in practical or applied philosophy. We attended five across the calendar year.[2] The overall experience was not encouraging. In most cases, concerns with developing a philosophical practice or a means for implementing ideas seemed marginal at best. There was little recognition of the basic difference between talking about socially relevant issues and getting involved in them. Plenary presentations provided accounts of philosophers exploring non-traditional spaces – for example, 'Investigating Discovery Practices: Studies of Bioengineering Sciences Labs'. But the narrative would then consist of a description and analysis of those aspects of the lab of potential interest to philosophers, rather than describing attempts to integrate philosophic insights into the work of the lab. Time and time again the conferences felt like exemplifications of our point about disciplinary capture.

One conference, however, left a different impression: the 2015 meeting of the International Society of Environmental Ethics (ISEE). One of us (Frodeman) has been attending environmental philosophy conferences since the 1990s, and had participated in the founding of an environmental philosophy association.[3] But it had been a half dozen years since I had attended any of these conferences. The ISEE meeting was held in Kiel, Germany, in July. The differences from the mid-2000s were striking. In the past, papers were read; now the majority of the participants gave talks illustrated by PowerPoint slides. Papers are by their nature more recondite, discouraging dialogue, while the talks are by their nature more conversational and accessible to a wider range of audiences (indeed, the ISEE audience was not limited to philosophers). This formed part of a more interdisciplinary atmosphere: where in the past one heard occasional references to real-world problems, usually in order to illustrate a philosophical point, presenters now demonstrated a significant understanding of the particularities of issues like climate change or the reintroduction of species. Tables and graphs presenting evidence for claims were common, and no longer elicited surprise (or alienation) on the part of the audience.

Across this six-year period the ISEE conference format had progressed from reading papers about, for example, the intricacies of intrinsic value or mentioning real-world policy issues in an off-hand way, to giving talks focused on real-world issues and in a few cases to actually becoming versed in the policy details of those issues. The shift was impressive. At the same time, there was little recognition of the need to go further: to offer an implementation plan for how these ideas can penetrate the larger world; or to actually get practically involved in these issues so that the philosophical ideas gained some traction; or to treat issues of implementation and impact as philosophic questions. People were now quoting the IPCC Reports, demonstrating a decent understanding of issues like carbon sinks, methane release, and problems with biofuels; but there were no theoretical or practical accounts of how their insights could reach stakeholders, via trickle-down, translation, or the co-production of knowledge.

We've noted what's missing in the literature – accounts of philosophical work done in the context of an ongoing environmental policy debate. But what would such an account look like? This is tricky, in part because doing justice to the context-sensitive minutiae that are so vital for policy-engaged work can take a book (and we have such a book, Briggle 2015). Nonetheless, we want to give a flavour of the kind of work, and of the kind of account of such work, that we have in mind. We'll switch to the 'I' pronoun here as this story is mostly about one of us (Briggle).

In October 2015, Denton Municipal Electric (DME, the public utility electricity provider for Denton, Texas) announced the Renewable Denton Plan (RDP). At that time, Denton got 30% of its electricity from a coal plant, 30% from purchases on the ERCOT market (Electric Reliability Council of Texas, the deregulated Texas grid), and 40% from wind power. The RDP proposed changing that mix to a total of 70% from renewables (52% wind, 17% solar, and 1% landfill gas), 17% from the market, and 13% from two new natural gas-fired power plants that it proposed to build in Denton. The cost to construct the power plants would be about $250 million. But the overall impact of the plan would actually reduce rates and lead to a $500 million cost savings over business as usual across twenty years. Environmentally, DME also projected that the RDP would reduce emissions of nitrous oxides, volatile organic compounds, particulate matter, sulphur dioxide, carbon dioxide, and methane by roughly 75%.

As a public utility, DME needed City Council approval to implement the RDP. The process also required a public hearing. In addition to those requirements, DME hosted two open house events and an open-mic Q&A event to answer public questions. The RDP met with strenuous resistance from a small but very active group of environmentalists who objected to the idea of building natural gas plants in Denton. DME has 40,000 customers and it's probably fair to say that 99% of the public engagement about the Plan came from less than 100 of those customers. The objections were about local air quality impacts, climate implications, and a general distrust of DME – indeed, most opponents of the Plan pushed hard for an independent auditor or third-party consultant to check the electric company's claims, a request that was eventually granted by a split 4:3 council decision.

This debate is set against the backdrop of Denton having just become the first city in Texas to ban the use of hydraulic fracturing to extract natural gas. There are 280 gas wells in the city limits of Denton and fracking was occurring at times less than 200 feet from homes. I helped lead the campaign for the ban. The group opposed to the RDP was largely composed of activists and other residents I had worked very closely with for over a year (in some cases five years) to lobby for the ban. Many of them expected me to lead the charge against the gas plants too, and they become frustrated when I took more of an open approach in my blogs and public comments.

I felt that my role was to help broker an honest conversation about the Plan, which was full of very difficult technical complexities, challenging values trade-offs, and conceptually confusing points. The fracking ban was about a land use policy issue of an industrial activity being sited too close to homes. This issue was related, clearly, but I thought it was different enough and complex enough for the appropriate posture to be one of openness and questioning rather than outright rejection.

As one example of conceptual complexity surrounding the issue, DME estimated that the renewable plan would reduce Denton's use of natural gas by 37%, which calculates to roughly twelve fewer gas wells required to serve the city's electricity demand. Many people thought this was smoke and mirrors: any construction of new fossil fuel infrastructure necessarily entailed more use of fossil fuels; these plants would only add more greenhouse gas (and other) emissions on top of existing emissions. After all, DME also calculated it would take fifty-eight gas wells worth of gas to feed the two proposed power plants. Yet I thought that DME's claims were conceptually sound. The gas used to feed those plants is gas that would have been used by other gas plants anyway. The proposed gas plants that formed part of the RDP would not just service 13% of Denton's needs; they would also produce electricity for sale on the ERCOT market. That electricity would displace electricity generated by older and more polluting gas plants that would otherwise be run on the market.

I increasingly found myself at odds with my (now perhaps former) friends. I felt that my role was to help elicit the strongest possible arguments both for and against the RDP. Much of the opposition was muddled and often attacking a straw man, so I tried my best, through blogs, conversations at City Hall, and in coffee shops, to separate the chaff of bad arguments from the wheat of good ones. Many people (including some from DME and the City Council) appreciated my efforts. Yet there were many who thought I was "vacillating" in a way that drained energy from "the mission at hand," which was to prevent the gas plants. Things got ugly, with some calling my blogs "vomit" and accusing me of being a shill for DME.

This is the other kind of 'hard' that we have talked about above. Acting as a philosopher in the field, I first had to listen to the claims being made on various sides and educate myself on the technicalities of the energy market. I set up extra meetings with DME officials, City Council members, and various residents who have different kinds of citizen expertise. In my early remarks I mostly called for more time to think and for more options than just RDP versus the status quo. I also criticized the process leading up to the RDP, which I thought had not been sufficiently inclusive or transparent. That process led to a great deal of mistrust and communication failures that might otherwise have been avoided. In my later remarks, including an op-ed in the local paper (Taylor and Briggle 2015) I questioned elements of the RDP and advocated for a wider range of options. Yet I continued to listen to proponents of the plan, and gained an increasing appreciation for its virtues, in large part by getting clear on some of the conceptually muddy points that had led me earlier to make some inaccurate claims.

Eventually, I honed my own position on the issue and shared my views with others, while inviting criticisms and counter-arguments. I came to think that many of the climate change-based arguments against the Plan were not very strong. I emphasized the need for air quality impact modelling for the natural gas plants – something the city didn't seem to emphasize until several of us started to push hard on this point. I also advocated for an incrementalist option of phasing in renewables slowly over time, as power purchase agreements appeared that made economic sense. Throughout my own advocacy I paid attention to my rhetoric. I emphasized that reasonable people could disagree on this issue because several values were at stake (rates, reliability, and renewables) that needed to be balanced, and because we were dealing with a thirty- or fifty-year timeline in a changing market. In such a situation the uncertainties and requisite interpretations could cut many ways. In this way, I hoped not only to foster a higher quality of democratic debate but also to model civility and humility.

If I were to turn to 'the public' to evaluate me, I'm sure (judging by Facebook posts) that I'd get a mixed review. Some called me "intellectually dishonest," "hypocritical," and accused me of knowing "zilch minus two" about the issue, espousing "neo-totalitarianism," and having my "chain yanked" by certain members of City Council. Others said things like (this from a City Council member about one of my blog posts) "this article lays out the key issues and key rubbing points between different perspectives better than anything I've seen." An interested citizen commented "for those of us who are still trying to get all the issues, I find this incredibly helpful. ... Thanks for sticking your neck out, and contributing meaningfully to the debate."

Make no mistake: my arguments have not carried the day. But I was able to add a bit more perspective to our community's conversation. My goals were to elicit more options and offer a model of civil, fair-minded discourse, while clarifying conceptual issues and drawing out hidden values dimensions – goals that easily transcended the specifics of the controversy.

NOTES

1. cf. http://plaisance-mclevey-lab.github.io/.

2. The five: Consortium for Socially Relevant Philosophy of/in Science and Engineering; Public Philosophy Network; Society for the Philosophy of Science in Practice; International Society for Environmental Ethics; American Society for Bioethics and the Humanities.

3. International Association for Environmental Philosophy, founded in 1997.

Chapter 5

Bioethics

As species of applied philosophy, environmental ethics and bioethics occupy different places in our social imagination. Environmental ethics strikes many as an arcane subject, of concern mostly to tree-huggers. The field has failed at its largest goal: convincing society that environmental health and sustainability are the sine qua non for all human endeavours. 'Bioethics', on the other hand, is intuitively understood. Possibly no other area of practical ethics is so readily grasped, and so quick to anger people, whether the subject is the rationing of healthcare, the high cost of prescriptions, or Obama's purported Death Panels.

There are also similarities between the two domains. As we've noted with environmental philosophy, medical ethics is also as old as Western philosophy. Moreover, in parallel with environmental ethics, bioethics is a creature of the twentieth century, its development spurred by historical events and technoscientific development. But we cannot claim, as we did with environmental ethics, that bioethics is a product of twentieth-century philosophy and disciplinary culture. When it emerged as a more or less distinct field in the 1970s, it was through the interdisciplinary efforts – sometimes competitive, sometimes cooperative – of lawyers, physicians, nurses, scientists, theologians, sociologists, and philosophers. Bioethics was never 'applied philosophy', if by that term we mean the application of academic philosophy to questions of social concern. Its provenance has not been so pure. And this has been a key to its greater degree of success in terms of influencing society.

In the West, medical ethics dates back to the Hippocratic Oath (fourth century, BCE), which codified physician responsibility (although this provenance has been questioned: parts of the Oath seem to reflect Pythagorean rather than Hippocratic philosophy). Questions of medical ethics show up in Plato: in the *Republic*, Socrates voices concern with excessive efforts to tend to the frail, and of course the platonic corpus as a whole is concerned with medical ethics, in that the philosopher is thought of as the physician of the soul. Moreover, both Plato and Aristotle discuss questions such as the permissibility of abortion and the appropriate level of care due to the body. Much more recently, the development of a formal medical ethics was a key part of the professionalization of medicine in the nineteenth century. The social power of physicians grew as they locked down a monopoly on the legitimate practice of medicine. Medical ethics, particularly as inscribed in codes of conduct, was the expression of the duty to wield that power responsibly. At the centre of that obligation was the principle of beneficence: do no harm. Medical ethics underpinned a relationship of trust: society relied upon physicians to use their knowledge for good rather than ill.

This trust was violated by the Nazi physicians who conducted heinous experiments on concentration camp prisoners. Following the war, international military courts put on trial twenty-three Nazi physicians, in what came to be known as the twelve 'Doctors' Trials' or the 'Subsequent Nuremberg Trials'. In an appeal to a principle transcending the laws of any one nation, the indictments referred to 'crimes against humanity.' One of the key outcomes of the trials was the 1946 Nuremberg Code, which sought to define legitimate medical research.

The central role played by scientists in the Nazi experiments highlighted the need to rethink the boundaries of medical ethics: research and therapy were increasingly interlinked, with each raising ethical and indeed metaphysical issues. Across the latter half of the twentieth century, the burgeoning powers of biomedical science and technology posed philosophical questions that challenged the parameters of the physician–patient relationship as well as the prescriptive status of the natural. 'Bioethics' marks the growth into a larger and more ambiguous field. It challenged the control physicians had exercised over the framing of the ethical issues at a time when patients were seeking greater autonomy and questioning authority. As two analysts of bioethics put it: "If traditional medical ethics was a form of self-critique and self-control, bioethics, at least initially, represented public critique and public control" (Dzur and Levin 2004, p. 336). The history of bioethics is in part a dispute over who this 'public' is and how that new creature, the 'bioethicist', should relate to that public.

BIOETHICS: AN INSTITUTIONAL HISTORY

We can date the contemporary institutional expressions of bioethics to World War II and the rapid proliferation of medical technologies across the latter half of the twentieth century. A quick look at this history highlights the interdisciplinary and transdisciplinary roots of bioethics, a field that did not arise from a single discipline within the academy, but rather one that accumulated like a mosaic from multiple sources, both practical and theoretical.

The Nuremberg Code appealed to the long-standing tradition of beneficence, but also[1] emphasized the principle of consent. Indeed, the code's first line reads, "The voluntary consent of the human subject is absolutely essential." The Nuremberg Code prompted the adoption of the Declaration of Geneva in 1948 by the World Medical Association (WMA). The WMA also adopted the Declaration of Helsinki in 1964, which outlined a set of ethical precepts and a guide to the protection of human rights in the conduct of experiments. These foundations of bioethics all shared a focus on personal autonomy.

Across the latter half of the twentieth century, emerging biomedical technologies raised a host of philosophical issues. Notable examples include the era of genomics and molecular biology unleashed by the 1953 discovery of the double helical structure of DNA, and the first successful human kidney transplant in 1954. The introduction of the birth control pill in 1960 and other assisted reproductive technologies (including in vitro fertilization, which, in 1979, led to the birth of the first 'test tube baby') set the stage for the sexual revolution. The 1960s and 1970s also witnessed "a dramatic shift from death at home to death in hospitals or other institutions" (Callahan 2004, p. 279). New technologies often gave physicians the power to produce effects that were not always beneficial for their patients. They could lead to the saving of lives that were severely limited, or force decisions that meant saving some lives at the expense of others (see Alexander 1962). "To save or let die?" became an overriding question, which was first systematically treated by one of the founding figures in bioethics, the philosopher and theologian James Childress, in his 1970 essay 'Who Shall Live When Not All Can Live?' Advances in palliative care made consideration of when to withhold or withdraw life-sustaining treatments an ever more pressing issue.

In the early 1960s, such questions lay at the centre of the renal dialysis debates. At that time there were not enough machines to serve public demand; the issue, largely faced at the state level in the United States, was how to distribute these scarce resources. Some states used a lottery system, but in Seattle, Washington, a committee of citizens was formed to make these decisions. Due to the gravity of its deliberations, *Life* magazine dubbed it the "God Squad" (Alexander 1962). In a 1964 speech, Dr Belding Scribner, inventor of the technique, predicted that the ethical problems his renal dialysis programme had encountered "will recur again and again as other new, complicated, expensive, life-saving techniques are developed."

In 1970 the biochemist and oncologist Van Rensselaer Potter coined the term "bioethics" in an essay that sought to bridge concerns with environmental and human health, as well as between the sciences and humanities (see Potter 1970 and 1971). A set of institutions began to form. In 1971, the politician Sargent Shriver, the Dutch priest and physician Andres Hellegers, and others at Georgetown University created the Joseph and Rose Kennedy Center for the Study of Human Reproduction and Bioethics, now known as the Kennedy Institute of Ethics. In 1969, the philosopher Daniel Callahan and the psychiatrist Willard Gaylin formed the Institute of Society, Ethics, and Life Sciences, later changed to the Hastings Center, which adopted an independent 'think tank' model rather than a university affiliation.

Throughout this time period, scandals in human subjects research continued to flare up. A whistle-blowing essay by anaesthesiologist and medical professor Henry Beecher in 1966, 'Ethics and Clinical Research', identified over twenty unsettling practices in human experimentation. Throughout many of the scandals the federal government had played a passive and ad hoc role. But in 1966, the U.S. National Institutes of Health (NIH) created a decentralized system of regulations that for the first time established collective supervision over decisions traditionally delegated to individual researchers. At the core of this system were review committees known as Institutional Review Boards (IRBs) – decentralized group boards composed partially of non-scientists that reviewed the ethics of research proposals involving human subjects.

In 1972, the infamous Tuskegee Syphilis Experiment was made public, where between 1932 and 1972 the U.S. Public Health Service allowed the natural progression of syphilis to go untreated in rural African-American men in Alabama. In the following year NIH recommendations encouraged the use of live foetuses for medical research before they died. In 1974, Congress passed the National Research Act, which prohibited federal funding for any research that violated NIH standards, including an amendment prohibiting research on a foetus with "a beating heart." Part of this public law mandated the creation of the temporary National Commission for the Protection of Human Subjects of Biomedical and Behavioural Research (aka the National Commission), chaired by Kenneth J. Ryan, MD. Congress charged it with identifying "basic ethical principles" to guide biomedical research involving human subjects. This included articulating the theoretical principles upon which previous regulations were based.

The National Commission was composed of eleven members drawn from 'the general public' and experts from scientific and non-scientific disciplines, including philosophy, law, and theology. No more than five of the members could be scientific researchers, indicating congressional commitment to bring human experimentation under outside scrutiny. The National Commission was charged with analysing issues of research involving human subjects. It produced reports on research involving vulnerable subjects including prisoners, those institutionalized as mentally infirm, foetuses, and children. It was the first federal-level US public bioethics institution, formally defined as "ethical inquiry conducted by a publicly constituted body, which is created and supported by government" (Fletcher and Miller 1996, p. 155). Though it had no enforcement powers of its own, the National Commission contributed to the first federal regulations for the protection of human subjects of biomedical and behavioural research. In 1979, the Department of Health, Education, and Welfare adopted these regulations, known as the Belmont Report (National Commission 1979), as a statement of its policy regarding the ethics of research involving human subjects. The regulations and principles outlined in the Belmont Report have since been institutionalized in the IRBs. The Belmont Report "had a major impact on the development of bioethics." Its principles found their way into the general literature of the field, and, in the process, grew from "the principles underlying the conduct of research into the basic principles of bioethics" (Jonsen 1998, p. 104). The principle of "respect for persons" has since become an important – and contested – centrepiece of bioethics.

Through this era, the concept of principlism became the leading approach in academic bioethics. At the same time a number of other approaches – including feminism, casuistry, communitarianism, and various theological and conservative traditions – grew alongside it and often competed with it and one another (e.g. Clement 1996; Callahan 2003; Breck and Breck 2005). Though principlism can be thought of as forming something like the 'mainstream' of bioethics, one could also see it as just one of numerous forms – virtue ethics, feminism, libertarianism, and religious ethics of all types – of bioethical inquiry. Across the rest of the twentieth century, bioethics grew in influence within public policy, public debate, the conduct of research, and the teaching of medicine. With one brief hiatus in the late 1980s, federal-level bioethics commissions worked at the highest level of government. In 1993, the US Office of Technology Assessment produced a comprehensive overview of US public bioethics (USOTA 1993). In addition to centralized federal committees, several other ad hoc, topic-specific advisory committees and decentralized local oversight and regulatory committees such as IRBs, Hospital Ethics Committees (HECs), and the Recombinant DNA Advisory Committee (RAC) fulfilled a number of oversight and monitoring functions.

Bioethics committees and ethics councils in other nations also multiplied, inspired by the success of the UK Warnock Committee (1982–1984) on assisted reproduction. Moreover, the 1980s and early 1990s witnessed the growth of alternative practices for anticipating and discussing the ethical and social dimensions of science and technology. This included the establishment in 1988 of the Ethical, Legal, and Societal Impacts (ELSI) programme. Chaired by Nancy Wexler, the ELSI programme was part of the US federal

Human Genome Project (HGP). As the principle agencies in the HGP, the Department of Energy (DOE) and the NIH devoted between 3% and 5% of the HGP budgets to the study of social, legal, and ethical issues (see Juengst 1996). Despite this largesse, constituting the best-funded philosophic research programme in history, the ELSI model pigeonholed ethics and values into a side stream and never systematically promoted the development of a fruitful dialogue with scientists or policymakers.

After the cloning of the sheep Dolly in 1996, President Clinton, a Democrat, requested a report from his bioethics commission on cloning. Their report recommended that federal regulation be enacted to ban research using somatic cell nuclear transfer (SCNT, i.e. cloning) to create children (NBAC 1997). Importantly, the justification for this ban was that cloning techniques were unsafe – a pragmatic rather than philosophical or religious stipulation. It was based on a certain weighting of the Belmont principles of beneficence and autonomy. It was this principlist approach to bioethics that Leon Kass criticized when he was appointed in 2001 by newly elected Republican president George W. Bush to chair the next incarnation of a federal bioethics committee. When Democratic president Barack Obama terminated Bush's committee and appointed his own bioethics committee in 2009, it rejected the style of bioethics practised by Kass – indicative of the ways in which bioethics tends to become implicated in US partisan politics.

Throughout the latter half of the twentieth century, institutes, centres, and departments of bioethics, medical ethics, and medical humanities rapidly multiplied (see Levine 2007). These institutions largely defined the field in terms of the ethical dimensions of medicine, healthcare, and biomedical science and technology rather than in the more sweeping ecological frame envisioned by Potter. There are now nearly forty master's programmes and six PhD programmes in bioethics. Though the field remains hard to define, by one reckoning there are over fifty academic journals devoted to bioethics. The annual conference of the American Society for Bioethics and the Humanities (ASBH) gives one a sense of just how large of an industry bioethics has become. Hundreds of members representing diverse academic fields and non-academic professions (philosophers represent about 15% of the total) gather to discuss ideas in hotel ballrooms lined with booths from dozens of graduate programmes, research institutes, and publishers.

A LOOK AT THE BIOETHICS LITERATURE

Once again with the help of our graduate student Kelli Barr, we reviewed a portion of the bioethics literature. Our criteria were the same: we were looking for instances of boundary work reflecting on the scope and aims of the field, and for accounts of integrating the insights of bioethics in society (reflections on impact), including descriptions of best practices. Due to the greater amount of literature in bioethics, we adopted a random sampling technique to survey five leading journals. After calculating average articles per issue, we estimated how many issues would need to be sampled to collect one-third of the total article population – giving us a sample size similar to applied philosophy and environmental ethics. For each journal, the individual issues were then numbered sequentially and a random number generator produced a number set corresponding to the number of unique issues to sample from that journal. We looked at the *Hastings Center Report* (founded in 1971), *Journal of Medicine and Philosophy* (1976), *Bioethics* (1987), *Kennedy Institute of Ethics Journal* (1991), and *American Journal of Bioethics* (2001). These journals have published 13,987 peer-reviewed articles; altogether, we examined 3,625 (see Table 5.1).

Once again the overwhelming percentage of papers included no accounts of researchers field testing their ideas or offering an implementation plan. Nonetheless, the percentage of meta-analyses – 4.2% – was triple the rate for applied philosophy (1.4%) and more than double of environmental ethics (1.75%) literature. Most of the meta-analytical literature (3.3%) amounted to boundary work, while only about one-fourth of the meta-analyses (0.9%) broached questions of success, impact, or best practices. Much of the boundary work in bioethics focused on questions of methods. In the *American Journal of Bioethics*, for example, our survey selection included three special issues on methodology: one focused on the role of narrative as a method for

Table 5.1 Our survey of the bioethics literature

	Article count	Date established (volumes surveyed)	Boundary work	Reflections on impact	Total meta-analyses
Hastings Center Report	1226	1971 (vols. 1–45)	24	12	36
American Journal of Bioethics	1163	2001 (vols. 1–15)	57	3	60
Bioethics	541	1987 (vols. 12–29)	14	4	18
Journal of Medicine & Philosophy	527	1976 (vols. 1–40)	15	11	26
Kennedy Institute of Ethics Journal	168	1991 (vols. 1–25)	9	4	13
Total	3625		119 (3.3%)	34 (0.9%)	153 (4.2%)

Source: Author's own

conducting bioethical case studies (see Chambers 2001), another on developing systems analysis in bioethics (see O'Malley, Calvert, and Dupré 2007; Robert 2007), and the third on the importance of incorporating empirical approaches into analytical frameworks (see Kon 2009).

A good example of reflections on impact and best practices comes from the *Hastings Center Report* (HCR), which featured a section titled 'Field Notes' from 2004 to 2008. These short pieces recounted experiences of working outside the academy. They were framed as bioethicists reporting back to 'base camp' from the 'field' of a hospital, policy advisory board meeting, congressional office, or medical school. Berlinger (2004), for example, describes the policy changes implemented at the National Council of Churches after her invited lecture on human genetics. While overseeing the Hastings Center's Public Interest Initiative, Johnston (2008) describes learning how to design talking points for reporters. Percentages aside, there was a discernible difference in outlook to the bioethics literature, a more practical tone and ease with the particularities of policy as compared with the applied philosophy and environmental ethics literature. The structure of HCR especially emphasized a different approach: articles were shorter, presented in the format of a magazine, and were intermingled with reports from the policy frontlines. HCR captured not only the state of research in bioethics, but also broader events surrounding policies and politics related to biomedical research and clinical practice.

What conclusions can be draw from these numbers? First, the question of impact is clearly more front and centre within bioethics. Something like one in twenty articles focuses on the topic, and a survey of entire articles suggests that it is more common within the bioethics literature for articles to integrate concerns with impact. Nonetheless, a case can still be made for disciplinary capture: most articles do not try to conceptualize, track, or evaluate impacts. For example, we could find little evidence of discussions about pathways to impact or mechanisms entailed in getting from thought to action. The discipline doing the capturing here, however, is not philosophy: it is rather the autochthonous field of bioethics that has established many of its own disciplinary features (e.g. journals and departments). In other words, bioethics was partially 'captured' by the prevailing disciplinary model of what counts as authoritative knowledge. Second, as noted in previous chapters, it's possible that the kind of reflexivity our study is designed to detect does indeed occur within bioethics, but not within the journal literature. For example, questions about the identities and roles of bioethicists and stories about practical successes in the field were quite present at the 2015 ASBH meeting in Houston, Texas.[2] It is apparent that members of that organization are keenly aware about the history of bioethics, its ongoing struggles to define itself, and its many institutional expressions that are inter- and transdisciplinary. We have yet to encounter a group within applied philosophy or environmental ethics as self-consciously reflexive as the bioethicists at ASBH.

Although there are books that reflect on identity and impact in these other fields, it certainly doesn't appear to be nearly as established as it is in bioethics. The kind of reflexivity we have in mind occurs in books such as Tristram Engelhardt Jr's *The Foundations of Bioethics* (1996), Kass's *Life, Liberty, and the Defense of Dignity: The Challenge for Bioethics* (2002), Judith Andre's *Bioethics as Practice* (2002), Jonathan Baron's *Against Bioethics* (2006), Jonathan Moreno's *Progress in Bioethics* (2011), and John Evans's *The History and Future of Bioethics* (2012). There is also an entire field devoted to the sociology of bioethics, which constitutes a tradition of reflection about bioethics and its place in society (see Sheehan and Dunn 2012).[3]

There's another explanation for the lack of reflexivity in these other literatures. In the cases of applied and environmental philosophy, publishing articles in journals is a poor way to achieve impact. Few outside of these sub-fields read these journals. Philosophers are then faced with a choice. They can hope that the passive diffusion model will do its work, or they can take the additional step of developing plans for integrating their insights into real-world problems. This should in turn lead to reflections about that work in the pages of peer-reviewed journals. In contrast, bioethics has become sufficiently integrated into various scientific and non-academic institutions that publishing articles in the top journals is in itself a way to reach wider audiences. So, for instance, an editorial in the *American Journal of Bioethics* (AJOB) was featured in the popular online magazine *Slate* (Mathis-Lilley 2015). Similarly, AJOB claims: "Our readers include faculty and students at virtually every graduate and professional school in the world, thousands of elected officials and judges, and most health news journalists." *The Hastings Center Report* claims that its readership includes "physicians, nurses, scholars of many stripes, administrators, social workers, health lawyers, and others." Bioethics journals may have become vehicles of direct and timely impact in their own right, rather than merely repositories of ideas.

THE SUCCESSES AND LIMITATIONS OF BIOETHICS

When we speak at conferences or other venues about how philosophy does not have sufficient impact, nor is adequately self-reflexive about how to have an impact, we can count on hearing the question: What about bioethics? We agree: bioethics is a significant exception to the general breakdown between supply and demand within applied ethics. Toulmin is largely right that "medicine saved the life of ethics" (1982).

Consider these three indicators of this success. First, it is easy to find examples of individual bioethicists working with stakeholders on topics of social relevance in real time. This not only includes people such as Arthur Caplan, Ruth Faden, Jonathan Moreno, Carl Elliott, and Leon Kass, who have received sustained attention in the popular press; it is also the case at ASBH, which has 1,800 members, many of whom are practising what might be called 'clinical bioethics', engaged work on particular problems in addition to or instead of disciplinary scholarship. Second, these are not just one-off, individual accomplishments (as is the case, say, with Andrew Light's involvement in environmental policymaking, or Peter Singer's impact coming out of applied ethics). Rather, bioethics has found ways to institutionalize practices of social engagement. Stable spaces for conducting engaged work include HECs (100% of hospitals with more than 400 beds have institutionalized healthcare ethics consultations), IRBs (found at every research university), bioethics courses regularly incorporated into the curriculum at medical schools, federal and state level bioethics committees, a department of bioethics at the NIH, and dozens of centres and institutes (mostly at universities). Even if the people working in these spaces do not self-identify as (full-time) bioethicists, the field is responsible for creating these spaces and shaping the moral discourse found therein (see Evans 2014).[4]

A third indicator is the existence of a reflexive literature within bioethics that critically appraises its identity and functions and raises questions of best practices and impacts. We noted above that we found roughly twice as many meta-analyses in the bioethics literature compared with the other literatures we surveyed. And we also suggested that in other ways (conferences, books, and sociological reflection) bioethics also stands out as a distinctively self-conscious field. One way to summarize these indicators of success would be to claim that bioethics has become its own profession. By this we mean that bioethicists have become members of a class that has the authority to speak on certain issues in the public sphere. Policymakers listen to them, and gate-keepers of discourse (such as editors of newspapers) give them a soapbox to help shape societal debates (see Evans 2014). While there is not a specific degree (like the JD or MD) for bioethics, there are numerous programmes offering bioethics degrees. And there is talk about a code of ethics for bioethicists (e.g. Baker 2007).

We see two reasons for the greater societal success of bioethics. First, it was born out of multiple disciplines – law, medicine, philosophy, theology, social science, and more. True, this created sometimes rancorous ongoing battles for dominance across these disciplines. But these interdisciplinary origins gave the field a robust mix of tools and perspectives from which to create a system of knowledge to legitimize its claims to authority. Bioethics took shape around boundary objects, such as kidney dialysis machines, that allowed for multiple fields to stake some turf along the borders. By contrast, environmental ethics and applied philosophy have consisted of a mostly one-sided push outward by just one discipline (philosophy) with little sustained and defining cross-pollination with other fields. In contrast, philosophers are a minority presence in the field of bioethics.

Second, recall the escape velocity metaphor and the failure of applied philosophy to break the bonds of disciplinary gravity. The founders of bioethics had a major boost: substantial parts of society (such as the U.S. Congress, and physicians) were asking for help. They had problems, such as scarcity of medical resources and mistreatment of human subjects of research, which they were defining as ethical in nature. They needed assistance from philosophers and others (see Callahan 1973). In other words, there was a 'demand side' to bioethics. By contrast, environmental ethics has mainly been 'supply side': while environmental problems certainly have ethical dimensions, they are often not framed as such – or the ethical dimensions have elicited an unhelpful 'owls versus humans' dynamic. Thus, the environmental ethicist has to push his or her perspective into often inhospitable terrain dominated by talk of economics, science, engineering, and even human rights. This is something we know first-hand when, for example, we give public talks about fracking or acid mine drainage and one of the first questions is often: "What does philosophy have to do with this?" It is not an impossible task to stake a claim to issues that has not yet been defined as philosophical in nature, but the practitioner must practice the art of the chameleon – spending long stretches of time blending into the surroundings and only standing out when the moment is right. The bioethicist, by comparison, usually does not need camouflage.

But if bioethics has been notably successful in terms of broader impact, we should balance that success against the difficulties the field has had. Increasingly, the literature of the field is sounding notes of apprehension (e.g. Eckenwiler and Cohn 2007). This literature is extensive, so that a comprehensive analysis is beyond our scope. We can, however, identify three overlapping kinds of concerns: of goods, methods, and legitimacy.

What we call the 'crisis of goods' is articulated by Judith Andre (2002). Andre conceives of bioethics as a practice. Like chess or football, there is a set of internal goods to bioethics, such as clarifying disputes surrounding values and assessing arguments; but there are also external and instrumental goods that get tangled up with bioethics, especially money, power, and fame. This sets the stage for a long-running criticism of bioethics: the problem of 'selling out'. This selling out takes a particular form. John Evans (2002) argued that mainstream bioethics obtained its influence by substituting a "formally rational" discourse for a "substantively rational" one. Bioethicists abandoned substantive debates about the good life, which allowed them to fit the classically liberal notion of instrumental expertise and harmonize with the basic goals of technoscientific market utilitarianism. This extends their social influence in innocuous ways. As Moreno (2005) put it, bioethicists struck a compromise. They get the social power that attaches to the label of 'moral expert,' and in exchange they forfeit any deep questions about the ongoing technoscientific remaking of the human condition. Bioethicists are given funding in exchange for guaranteeing "science a green light disguised as a flashing yellow" (Moreno 2005, p. 14).

For Evans, bioethicists pass off as a neutral expertise what is in reality a contingent way of framing moral problems. They have constrained dialogue and narrowed the available options without opening up these constraints for debate. It has had an exclusionary effect on any concerns that do not fit the supposedly neutral mould of the discourse. What, for example, are we to make of claims about humans overreaching their proper sphere of action, to 'play God', or about the deeper anthropological and philosophical meaning of procreation versus making? These questions are bracketed out of the conversation insofar as the bioethicist (acting as a moral expert) prefigures the discussion in principlist terms. The principlist framing is one that seeks to balance rights (autonomy), risks (beneficence), and access (justice). Insofar as these constitute the only legitimate values, questions about, say, human dignity are treated as the irrational musings of amateurs. This amounts to a narrowing or thinning of ethical discourse, one that is aligned with the interests of money, science, and power (Briggle 2010).

But the issue, rather than simply being one of 'selling out', also raises questions of rhetoric. In 'Bioethics as a Discipline', Daniel Callahan emphasized that bioethics should be something useful to those who face real-world problems: "the discipline of bioethics should be so designed, and its practitioners so trained, that it will directly – at whatever cost to disciplinary elegance – serve those physicians and biologists whose position demands that they make the practical decisions" (2007 [1973], p. 21). Callahan argued that ethicists had two choices when it came to addressing their difficulties in communicating with scientists and physicians. They could "stick to traditional notions of philosophical ... rigor" and continue "to mutter about the denseness and inanity" of their non-philosophical colleagues, or they could adapt an extra-disciplinary definition of rigour:

> Not the adaptation of expediency or passivity in the face of careless thinking, but rather ... the kind of rigor required for bioethics may be of a different sort than that normally required for the traditional philosophical or scientific disciplines. (Callahan 2007 [1973], p. 19)

As an example of a new kind of rigour, Callahan claimed that a good bioethical method would allow those who employ it to arrive at quick and viable solutions to moral dilemmas. If bioethicists are to be of use in such situations, philosophical reflection must be tempered by the demands of timeliness. They must have ready at hand a "normative ethic, which can presuppose some commonly shared principles" (2007 [1973], p. 21). Otherwise, bioethicists will remain mired in foundational debates restricted to an insular set of colleagues that would seldom contribute to actual decisions in real time. The principlist framework of the Belmont Report allowed for bioethicists to offer real-time answers, the very thing Callahan knew would be needed for the field to succeed. Yet providing answers often came at the price of predefining and thus narrowing the range of inquiry. The question thus becomes: how can you give timely advice while still staying attuned to the richness of a topic and the diverse perspectives that can be brought to bear on it?

The second concern could be loosely described as one of methods, but really raises core questions of identity and purpose. Of course, every profession has its theoreticians who argue about what kind of principles or methodology underlies their work. In this way, bioethics is no exception. The field features a long-standing debate between top-down principlism and bottom-up casuistry as well as voices from pragmatism, feminism, and more. The difference is that in bioethics such debates are so deep and intractable that it has led some to ask, "Does bioethics exist?" (Turner 2009). Can a postmodern queer theorist and a Baptist theologian really form parts of the same field? Perhaps the gulf between casuists, principlists, religious ethicists, and feminists is so wide that there is nothing that really holds them together as members of the same field or profession. Of course you can get two lawyers to offer contradictory positions (indeed, that's essential to the profession). But their modes of argument will be the same and they will draw from a common tradition, a common stock of theoretical knowledge, and a fairly standardized educational background and accrediting system. A feminist and a principlist may not even share this common ground. All they have is the boundary object under consideration (e.g. cloning or stem cell research). The question is whether that is strong enough to hold a field together such that those engaged with it could be given the same label and occupy the same social role.

The third concern, that of legitimacy, is the consequence of the first two concerns. Evans (2014) argues that bioethics obtained its credibility from "jurisdiction-givers" (the Congress, physicians, media, scientists, patients, and hospitals) by developing a common morality that allowed them to clarify values rather than pronounce on what someone's values should be. Bioethicists, he argues, found a way to be the neutral arbiters of others' values. This established their jurisdiction in the realm of hospital ethics and research ethics, where bioethicists are still largely seen as clarifying and defending shared values. Things got shaky, though, when bioethics attempted to extend its jurisdiction into the wider arenas of policy decisions and cultural debates, especially about matters of the beginning and end of human life. Here, it is much harder to identify common ground. A chasm divides conservatives and liberals on such issues as abortion. The same is true in terms of debates about transhumanism, the doubling (or more) of the human life span and proposals for enhancing humans either physically or cognitively. This has led to a growing clash between the proactionary embrace of the transformative power of science and the precautionary scepticism of it (e.g. Fuller 2011). Here too the profession of bioethics cannot speak in a unified voice, which undermines its ability to speak authoritatively. This has caused Edmund Pellegrino to note the "diminishing confidence" with which "the public seeks assistance" of the "bioethics community" due to multiple opinions (2006, p. 576). It begins to look like there are conservative and liberal bioethicists who can give you an ethical justification for any foregone political position. If that's the case, then what's the difference between a bioethicist and a spokesperson for the National Right to Life Committee or Planned Parenthood?

Such politicization naturally erodes claims to legitimacy based on offering a type of knowledge that is not reducible to special interest politics. Thus, as Evans sees it, the true enemy of bioethics is social-movement activism. Groups speaking for only certain segments of the population will colonize territories that bioethics had tried to claim in the name of some common ethical framework. Daniel Callahan has concluded that "the general public, and the medical and health policy world, will find it all too easy to dismiss bioethics as ideology driven, left or right politics in sheep's clothing. If we besmirch each other long enough, the public will soon conclude that we are all frauds" (2005, p. 431). There is a catch-22 at work here, because one could argue that the real crisis of bioethics is not its failure to represent universal values but the fact that no such things exist in the first place (something that Engelhardt 1996 has long argued). Perhaps Evans has misunderstood the role of bioethicists as neutral arbiters of values. It could be that their role, at least in broader policy debates, is to make the strongest possible case for marginalized and apparently weak positions. They would, then, serve the common good by cultivating a diverse ecosystem of ideas. Evans doesn't seem to disagree with this. But his point is that bioethicists can only effectively make these arguments if they are able to utilize their positions of institutional privilege. And the legitimacy that comes from occupying these positions derives from a general sense that being a bioethicist is something different than being just another run-of-the-mill activist or special interest group representative. Maybe bioethicists can speak authoritatively in the name of certain segments of the population or on particular values, but they've got to do so in a way that doesn't smell like the self-serving demands of special interest politics.

This may be the balance that any socially relevant philosophy must walk. Tip one way and you are dismissed as an activist. Tip the other way and you are ignored in your abstract theoretical quest for universal values. You need to find the right kind of rigour, something between activist rants and specialist curlicues. This calls for greater attention to be paid to the rhetorical dimensions of philosophic analysis.

CONCLUSION

Part II (chapters 3, 4 and 5), surveyed three responses to the general problem of philosophical irrelevance. We've argued that applied ethics and environmental ethics did not frame this problem appropriately, as one rooted in the institutional imperatives of disciplinarity. Bioethics, we've claimed, presents a different case, because whether by accident or design it has been more aware of and thus able to escape disciplinary capture.

What philosophers for a hundred years have taken to be philosophy tout court is in fact just one variety of philosophy. Nor is the problem only within philosophy: the disciplinary model generally does a good job of concealing its own contingency as only one way of organizing knowledge. In establishing internally policed standards it rejects every other approach, treating other modes of organization not as alternative models but as politically tainted or as consisting of mere amateurism. And once you accept the disciplinary standard, then by definition nothing else can count as rigorous, properly vetted knowledge – whether in philosophy or any other field.

The self-concealment of the disciplinary model explains why it was so easily embraced by applied philosophers and environmental ethicists. They wanted to speak to the issues of the day, but they could not see the infra-epistemology or the subterranean politics of disciplinarity – that this medium was more important than any message it might carry. Yes, occasionally individuals have broken free and gained traction in the wider world, but they remained one-offs, accidents in orbit around an unconscious dogmatism. The nature and paths of extra-disciplinary success have never been institutionalized or brought to self-awareness.

Bioethics is a different creature in at least three ways. First, it has an inter-disciplinary and transdisciplinary origin story. There was never *a* model as the assumed matrix of thought. Bioethicists had to work out how to speak with a variety of stakeholders. This was understood as an ongoing struggle: there is plenty of self-criticism and anxiety about the hands-on nature of bio-ethics, but it was generally recognized that it is better to learn on the fly than to offer nostrums from the sidelines. The second way bioethics stands out is in terms of the demand-side pull from society. Policymakers, doctors, the media, and others were asking for help with problems that they recognized as partly ethical in nature. This led to a third point: rather than being housed in a discipline, bioethics became institutionalized both within the academy and beyond it. A wide range of boundary organizations – IRBs at universities, HECs, and think tanks like the Hastings Center, in addition to professors housed within universities – kept standards for research fluid and responsive.

In part III, we draw inspiration from these aspects of bioethics to introduce field philosophy, our own alternative to the disciplinary model. Field philosophy is not limited to the topics native to bioethics (e.g., cloning) but can roam freely wherever important issues are being discussed. It is a way for philosophers to have more direct and varied impacts in real time – which raises a host of questions about the impacts of philosophy and the philosophy of impacts.

NOTES

1. This section draws from Briggle, Adam (2010). *A Rich Bioethics*. Notre Dame, IN: University of Notre Dame Press.

2. ASBH was founded in 1998 through the consolidation of three existing associations. It has over 1,800 members, making it one of the largest bioethics organizations.

3. A Google search for 'sociology of bioethics' returned 24,700 results whereas the same search for 'sociology of environmental philosophy' returned just one hit and 'sociology of applied philosophy' had zero hits.

4. It's worth noting that Congress, via the America Competes Act of 2007/2015, is mandating graduate instruction in ethics and values at every institution that receives public funding.

Interlude 2

Philosophical Spaces

We argued (in part I) that philosophy has suffered from being housed within the disciplinary structure of the modern research university. Since philosophy saturates all facets of our lives we need philosophical institutions out and about in society. Philosophy certainly belongs within the university; but it should not exclusively (or even primarily) function as a discipline: it is both more fundamental and more interstitial than that. Philosophers should circulate through the disciplines, seconded to other departments, tasked with a project for a period of time, before returning to their disciplinary home to recharge their batteries. And they should go abroad in the world, living for a time, and perhaps for good, in other institutional structures in both the public and private sectors. At the very least, the field needs to become more aware of the philosophical costs of disciplinarity, and should take steps to mitigate those costs.

In part II, we surveyed attempted solutions to the problem of the social irrelevance of philosophy by examining three areas of applied philosophy. We argued that despite a number of successes – most notably, but certainly not exclusively, in the field of bioethics – these fields suffer from disciplinary capture. We have further noted that much of the success in bioethics depends on the existence of a 'demand side' for philosophic work. Now, in part III (chapters 6 and 7), we offer our own way forward for the problem of impact – a model that we call field philosophy. We then sketch one element of the future research programme this model implies: the development of a philosophy of impact.

Let's say you are a philosopher in a philosophy department and no one is beating down your door with demands for you to philosophize about their problems. Yet you want to be more relevant. You want to have an impact. What can you do? One option is to rely on the disciplinary model of impact, the passive-diffusion, trickle-down model that everyone assumes. Effects may be quite indirect and occur over the long term, but you do have the security of having your work measured by standard bibliometrics like citation counts. Just crank out publications in peer-reviewed journals and hope one of them catches fire. Or you could extend that model, by sending your published manuscript to people in the public or private sectors who deal with the issues you write about. You could also push things one step further and arrange a meeting with them to help give your ideas some traction in their world. Now you are starting to do rather unorthodox things. So you'll need alternative ways of accounting for the impact of your work. The people you speak with may never cite you in an academic journal, but they might pass your ideas along in other ways. They might even put them to the test by implementing them.

Then again, you don't have to begin with the disciplinary model at all. You can set up a meeting with a city council member to see what kinds of issues they are struggling with. Maybe there's something philosophical buried within the issue, unacknowledged; in our experience, there always is. There are as many starting places as there are issues being discussed and decisions being made. Initially, you're just listening and trying to figure out if there's a project that falls into the sweet spot of the Venn diagram where your skills and passions overlap with a societal need. With a little creativity you are likely to find a space where you can have a role to play. We've heard complaints that such an approach sacrifices one's autonomy. But insisting on starting with your own concerns isn't autonomy; it's academic narcissism.

For us, these are the beginning moves in what we call a field model for philosophy. It can take a million forms, but its heart lies in its audience (stakeholders outside of the discipline) and in its beginning – not in the lonely office but in the hurly-burly. It is a way of listening and then a way of looking – looking for those moments when you can make artful contributions to help clarify a conceptual muddle (or muddle a false clarity), excavate and evaluate hidden assumptions, articulate and critique an argument, foster a more civil and informed dialogue, play the devil's advocate, or any number of the other things that philosophers habitually do. It will also almost always require your learning new things far from your specialty and working with people with diverse perspectives and backgrounds. (For us, this is part of the fun of it.) If you become a field philosopher, your effort may not be widely understood as a 'philosophy project' at all, and you may only publish a peer-reviewed article after the dust has settled. But that's okay, because that won't be the only or even the most important measure of your work. You'll be out there making waves in the spaces between the academy and the informal Republic of Letters, where philosophy is happening in real time.

What does field philosophy look like? We gave one example at the end of chapter 4 with regard to an energy policy proposal at the municipal level.

Another example began when one of us (Frodeman) served as a reviewer on a NSF panel in 2001. Listening with philosophical ears paid off by revealing a conceptual confusion. The reviewers were tasked with evaluating research proposals by two criteria: intellectual merit and broader impacts. Researchers from different disciplines was meant by 'broader impacts'? And how did it relate to intellectual merit?

CSID impacts, 2008–2011

Figure interlude 2.1 A rough map of the impacts resulting from our research grant "A Comparative Assessment of Peer Review" (CAPR). *Source*: Author's own.

This experience led us to put together our own research proposal, which we called the Comparative Assessment of Peer Review (CAPR). This four-year (2008–2012) project, funded by the NSF's SciSIP programme, examined the peer-review processes at six science agencies around the world. We sought to build a taxonomy explaining how different agencies conceptualize broader societal impact issues and integrate them into the ex ante review of grant proposals. CAPR's products included ten peer-reviewed publications examining the development of peer review at federal agencies worldwide. We also held a series of workshops with user groups in the US and Europe to disseminate and to field-test our findings. We conducted a survey of stakeholders concerning the relationship between science, society, and the role of peer review. And we contacted both the NSF and the US Congress about our work. The graph above attempts to get a handle on our impacts.

Of course, these arrows conceal as much as they reveal. For example, our meetings with NSF representatives and policymakers didn't simply follow from our publication. That took extra effort – the kinds of things we didn't do when we worked on the Nantucket Sound wind farm issue. This meant getting on the phone and on email with the audiences we wanted to influence, wheedling invitations to meetings and the like. We acknowledge as well the problem of *post hoc ergo propter hoc*: that just because an event follows upon our work doesn't mean that it was *caused* by our work. Events are multicausal: if we had an impact, our work was unlikely to be the only factor involved. Finally, we note what could be called the 'Freudian' aspects of impact: people do not always know when they have been impacted; and when asked, they might well deny it for any number of reasons.

In short, we know that this diagram is insufficient, but we view it as a start at charting (and thinking about) impact. It's an example of the kind of meta-philosophy or philosophy of impact that we had hoped to find in the literature. In our work, trying to understand the impacts of field philosophy (the focus of chapter 6) drives us to questions surrounding the philosophy of impact (the focus of chapter 7).

Part III

REACHING ESCAPE VELOCITY

Chapter 6

Field Philosophy

1. AN INVERTED WORLD

This book has multiple origins. One dates from 1985, when one of us (Frodeman) attended his first philosophy conference. The norms of the conference came as a surprise. Presenters stood at a podium, head down, reading from a text for thirty or forty minutes. There were no interruptions; questions or comments were held to the end. The papers were complex, written to be read rather than listened to. If you paused to think about a point, you missed a paragraph. Think a couple of times more and the thread was lost.

Philosophy is existential, dangerous, erotic. But the presentations were studied and abstract, disembodied and passionless. Where was the laughter, the playfulness? The stories relating the great ideas of the tradition to everyday life? And the venue was all wrong. It was a sensory deprivation chamber: ugly rooms at a nondescript hotel, no shades or shadows, no colours or windows, nothing green or growing, nothing to suggest that philosophy was exciting, current, or colourful.

Things were different that evening. People no longer dressed like IBM salesmen. Over drinks the playfulness returned. Tables were moved, chairs pulled around. The excitement level rose. People laughed, argued, flirted. Noting the difference in topics, interaction, and energy level, I asked my companions: Why can't we (sans the drinks, perhaps) bring the approach of the evening into the events of the day? Why not have ten-minute presentations, workshops, or charettes? Or couldn't we simply *talk*? This elicited laughter. I persisted: isn't the form and venue of our talks a topic for philosophical reflection, too? I received no response.

Over the next few years I tried to conform. Then, hesitantly, innovate. I included personal or literary elements in my papers. I received blank stares. One colleague announced that he would never again sit on a panel with me.

Occasionally something different occurred. In 1998 Alphonso Lingis was the plenary speaker at an environmental philosophy conference in Denver. He asked the moderator to turn the lights off before he entered the room. He then slipped into the darkened room by flashlight which, hung from his neck, created weird flickering shadows across his face. He had covered his face in gold glitter, and had placed a boom box behind the podium. He turned on a tape of crickets, animal cries, and the sounds of weather. Then he read his paper.

At the cash bar afterward I asked people what they thought of the talk. Reaction was evenly divided. Half thought the mise en scene was a wonderful lark; half found it disgraceful. But no one suggested that there was a philosophic point to the performance – that the light and glitter and jungle sounds were meant to evoke a campfire amid a natural environment, perhaps the original evolutionary environment of our species, as commentary on and counterpoint to the narrative concerning our place in the world. The performance was taken as mere spectacle.

I wrote my dissertation under Lingis. The presentation in Denver was of a piece with his overall approach to philosophy. Lingis had worked his academic schedule so that it consisted of six months teaching at Penn State and six months overseas. In the 1980s I would occasionally drop him off at the airport at the beginning of these six-month excursions; in the days before laptops, he would pack up a portable typewriter and two huge suitcases of books and head out to Managua, Port Moresby, or Bangkok. His days overseas were divided in half: one part spent reading Maimonides or Kant, Aristotle or de Sade; the other part spent wandering – visiting temples, strip joints, and the poorest parts of town. The texts that resulted were often jarring. In works like *Excesses* he chronicled the life of the mind brought into confrontation with the world of practical exigencies. Ideas were tested, sometimes at some personal risk. The results were at turns shocking and tender. It was philosophy that took its placement in the world seriously.

2. PHILOSOPHY'S MATERIAL CULTURE

It is not only the theoretical assumptions of the discipline that marginalize experiments like those of Alphonso Lingis. It is also the social and material culture of the university. Philosophers inhabit individual offices – or, more commonly these days work at home. You may ask "What else?" And we most certainly want to defend the need for places for contemplation. But philosophers need to vary their material culture.

Change could begin in the department. Philosophers need common spaces – not simply a lounge but a lab space where ideas and projects can be worked on together. "But there *are* no common projects." Yes, exactly. Creating the space will affect the kind of work that philosophy takes on – what and how things get written. "But the space will sit unused: people won't stop doing what they've always done, writing single-authored papers." True enough, until we change the incentives of philosophic culture – *and* attract a wider range of personalities into the profession.

Single-authored publications are the gold standard of philosophy, and co-authoring is frowned upon – in the words of one of our faculty evaluations, co-authoring "suggests a lack of research creativity." The result is particularly striking when philosophers produce communitarian treatises from within organizational structures, practices, and habits of mind that are so thoroughly libertarian in nature. The attitude is deeply engrained: as one of our colleagues puts it, "We are not really a department; we're a group of individuals who come together occasionally for departmental functions." But things can be nudged in other directions. Change the incentive structures of hiring, pay, tenure, and promotion; support collaboration and co-production; and find projects with outside groups that require philosophical skill sets.

Of course some will ignore the new opportunities. That's fine; we don't want a monoculture. But others will adapt, for idealistic reasons, for course releases, or for pay bumps. For these people, a common workspace will prompt conversations where suggestions can be made and connections explored. And move a couple of philosophers into the chemistry or economics department. Incentivize them so that they commit to spending a couple of hours a week in someone's lab. Conversations will develop. A discussion might prompt a new insight in one's own research – and perhaps a new idea for the chemist's research. Slowly, a new sense of what it means to philosophize will develop.

Today the interlocking elements of our institutional life create a supportive web for solitary habits. The campus is an enclave set off from the community, just as on campus philosophers are ghettoized from our colleagues in the other disciplines. Instead, open a storefront in a mall where philosophers make themselves available to the public. Have philosophers inhabit that space for a few hours a week, and change tenure standards so that they are rewarded for writing about the experience – in the local paper, no less. Place a graduate student in an internship at the EPA or in the state capital, making sure that they do not simply hide in an office but become part of the flow of things. Have them author 800- and 1500-word pieces for their new colleagues on the relevance of what they are doing to the situation they find themselves in. Over time their sense of the philosophic self will shift.

Today, philosophers have closer contact and deeper commitments with disciplinary colleagues around the world than with their extra-disciplinary co-workers across campus. Traditionally tied to campus by the need for library access, they now appear on campus twice a week, for classes, office hours, and the occasional meeting. They work hard, but on their own. The Internet has made it possible to access library materials while sitting in Copenhagen: one can now work anywhere, via laptop and wireless connection, doing research and delivering video lectures to classrooms back home from anywhere around the world. Take advantage of this fact: create partnerships with other universities, visit locales relevant to class material, and teach remotely, sending in reports from the field.

Field practices of philosophy can take as many forms as there are venues for thinking. For instance, philosophers could take a page from the history of land grant universities. Beginning with the Morrill Act of 1862, land grant universities created agriculture extension stations in every state and county. As part of this system, researchers in agriculture and home economics went out into the community to see the challenges faced by farmers and ranchers. Today there is a wide range of issues (the use of pesticides, GMOs, questions of social justice) tied to agriculture. Why are there no philosophy extension agents?[1]

We noted in the last chapter that philosophers have always, usually tacitly, imagined their work as effective, at least over the long term. They have implicitly and on rare occasions explicitly embraced the passive diffusion model, where ideas slowly make their way into the larger world. Field philosophy and other Mode 2 practices complement the diffusion model with an engaged approach that involves audiences in a direct and ongoing fashion.

Of course, with some exceptions, the world does not realize or even welcome the help of philosophers. We're seen as gumming up the works. Part of the challenge, then, is finding ways to increase what Boardman (2014) has called the "absorptive capacity" of individuals and institutions for philosophic perspectives. This implies a willingness to learn about different fields, the need to give greater attention to questions of rhetoric, and the ability to create philosophic boundary organizations that mediate between disciplinary philosophy and society.

Nor should our focus only be on research. Take an example from teaching. What if we left behind the curricular categories of logic, epistemology, metaphysics, ethics, and aesthetics, and instead taught courses the titles of which changed each semester? Teaching courses on Twitter, the Zika virus, El Nino, and the next election, where we show students how to approach these various events from a philosophical point of view? This doesn't mean *abandoning*, for example, metaphysics; it means *changing the direction* from which we come at these traditional questions. Parfit's discussions of personal identity are still relevant, but now in terms of Alzheimer's or Second Life. Students thus learn how to see the perennial questions of philosophy in the events of contemporary life. And they would get a sense of how philosophy appears in a variety of guises – that an issue like climate change raises a variety of ethical, epistemological, aesthetic (etc.) questions.

Professors get a bad rap: in the main they work hard. But whatever their personal politics they embody the libertarian self, assuming that they should be able to do as they please when they please. This extends to their choice of research: we've had colleagues react with horror at the idea that they should look for a research topic that would be of interest to an outside group. Begin from external motivations? That would impinge upon their autonomy! But autonomy can present as obliviousness, or narcissism, as tenured professors ignore the seismic shifts the university is undergoing: funding drying up, academic freedoms eroded, and a caste system taking root as universities increasingly rely on contingent labour to teach 'consumers'. Aren't we clever enough to find something interesting within almost any societal problem?

The solution is not to eliminate tenure – which, alas, has been in decline for forty years. Society needs a class of people with the time, security – and yes, leisure – to think new, provocative, and unconventional thoughts. But our habits and incentives need to change. Curiosity-driven research deserves support, but not when it becomes an excuse for indulgence or an exclusive focus on issues of interest to only a coterie of disciplinary connoisseurs. The point is a pluralist one: not to eliminate the status quo, but supplement it, treating philosophy as simultaneously disciplinary, interdisciplinary, and transdisciplinary in nature. Create space for all of these types within a department, and treat these modes as complementary. And encourage people to circulate through the three roles, giving them time to recharge their batteries after a period in the field.

3. THE DISCIPLINARY UNIVERSITY AND THE REPUBLIC OF LETTERS

The disconnection between philosophy and society is partly the result of the choice of which science the philosophic community views as paradigmatic. In the nineteenth-century geology held a central place in the cultural imagination. But across the twentieth century, as the philosophy of science became "philosophy enough" (Quine), philosophers treated theoretical physics and laboratory science as science *uberhaupt*. We discovered years ago that both scientists *and* philosophers viewed field sciences like geology as poor approximations of what properly takes place in the lab (Frodeman 1995; this earlier work on field science is one of the sources of field philosophy). The truth of the matter is the reverse: lab sciences create an artificially purified world of 'facts' that are of dubious worth once they enter the promiscuity of everyday life. Like disciplinarity itself, the strict border between 'inside' and 'outside' that characterizes the lab is an anachronism from the era of atoms and nation states in an age dominated by bits and transnational corporations.

The status of lab science is an iconic feature of the modern research university. It serves as theoretical justification for the disciplinary structure of the university, embodying the modernist dream of a certain knowledge outside of and dictating to politics. The modern university – whether we date its inception at 1810, with the creation of the University of Berlin, or 1876, with the founding of Johns Hopkins University – was itself an epistemic as well as political response to changing cultural conditions. Medieval universities began as private corporations of teachers and their pupils, bound together as a whole (*universitas*). Professors lectured (from the Latin *legere* 'to read'), a necessity in an age of scarce books. The study of the trivium and quadrivium allowed the students to master the accumulated insights of Western culture. Innovations (i.e. research) did not form a major part of the scholar's remit, as they were likely to arouse suspicions of heresy.

As Wellmon (2013) argues, the stresses now being felt by the modern university have a precedent in the crisis facing the medieval university at the end of the eighteenth century. Sixteenth-century Europe saw the rise of a parallel institution of knowledge to the university, the Republic of Letters. This informal network of intellectual activity brought together a creative scholarly and literary community that increasingly became the centre of intellectual life. A river of letters, first in Latin, later in French and other vernaculars, flowed between individuals – some unaffiliated, some housed at court, others at universities – that also connected institutions such as the Royal Society in the United Kingdom with salons in France and America (Grafton 2009).

The existence of the Republic of Letters challenged the status of the university scholar, who was increasingly viewed as a useless pedant who did not know what to do with the learning he possessed. Moreover, this alternative structure developed alongside steady improvements in printing technologies and the growth of a bourgeois reading public. By 1800, Europe was inundated by a "plague of books" that undercut a central justification for the medieval university (Wellmon 2013, p. 4). Society was suffering from an information overload – by the standards of the age – at the same time that the university was increasingly viewed as a place of pointless erudition, student drinking, and duels. How were individuals and institutions to navigate this wealth of knowledge, and tell what was trustworthy or authoritative?

A similar story played out within American higher education sixty years later. By the end of the Civil War, the traditional American college had become an anachronism, its role of conserving knowledge challenged by the increasing availability of knowledge via printed matter, and the growing need for people with high-level technical skills rather than simply a gentleman's education. The new universities – Johns Hopkins, Cornell, the University of Michigan, and a reconstituted Harvard – responded to these needs by embracing a new role: they became the institutions that would both produce and certify knowledge. They did this through an innovation in knowledge management: the development of the discipline.

The disciplinary university divided knowledge into discrete packets. This remains the sine qua non of authoritative knowledge. The notion of authoritative knowledge, that is, expertise, was itself dependent upon an ontological presumption: that it is possible to divide knowledge into units (disciplines) that are in principle separate from one another. This constituted an ontology of external relations where individual items are understood prior to their relations to other things. This is in contrast to an ontology of internal relations where individual things are constituted *by* their relations to other things. In the latter case the world is seen as being all tied together: in the words of John Muir, "When we try to pick out anything by itself, we find it hitched to everything else in the Universe."

Note too the political function of disciplinarity. By placing knowledge production in an ideal space, a premium could be placed on epistemic certainty. Certainty is only possible when outside forces are no longer in play and where all the remaining variables can be controlled. Experiments repeated under identical conditions give identical results. The resulting certain knowledge can then be delivered over to society as a set of inescapable facts that society must acquiesce to. While the use of those innovations are politically charged, their origin is pure, uncontaminated by values or bias. Politics, understood as merely the clash of subjective preferences, is thus bound by certain agreed upon truths. This, however, is a politics devoid of philosophy, a politics of subjective preferences.

While the amount of additional knowledge is unimaginably greater, our situation today approximates that of the beginning of the nineteenth century: a knowledge culture under severe stress. The limitations of lab science, the crisis of the modern research university, and the challenges facing philosophy all highlight our increasing reliance on non-disciplinary ways to manage knowledge. The certainty of the lab, or more recently the virtual 'lab' of the computer model, is giving way in an era where our problems are irretrievably interdisciplinary in nature. The university's long run as *the* modernist institution of knowledge production, providing a privileged, apolitical, and a-philosophical knowledge is at an end. Of course Latour is correct that we have never really been modern, that the separation of knowledge into disciplinary warrens was never truly possible. But the attempt was plausible enough to carry us along for quite some time. That plausibility has been lost: we are entering an era where we can't simply appeal to 'the facts'. And unless we are willing to give in to utter subjectivism (alas, increasingly the state of political discussion, aka politics as a fact-free zone), we will be driven to philosophize.

This epistemic and political unruliness is giving birth to a new set of knowledge institutions (one sign of this: the top ten companies worldwide, from Volkswagen to Merck, spent more than \$100 billion on research in 2013 [Casey and Hackett 2014]). A new Republic of Letters has formed – extra-university sites of knowledge production, whose boundaries are permeable and ever-shifting. Sanjena Sathian notes this development in her 2016 *OZY* piece 'The 21st Century Philosophers'. She argues that the entrepreneurs and technologists of Silicon Valley and other hubs of innovation today function as de facto philosophers. Sometimes they work on the margins of the academy in units like Oxford's Future of Humanity Institute. But more often they are located in non-academic locations like the Center for Applied Rationality, Google, and the Breakthrough Institute – or at magazines (e.g. *Wired*), blogs large and small, YouTube channels, and other social media. These are the modern-day salons – though they make up a decidedly uneven landscape, in some cases backed by enormous amounts of capital that are able to turn ideas into realities, with all the profits and problems that follow.

There are notable differences between these de facto philosophers working in the contemporary Republic of Letters and their academic brethren. First, some are writing best-sellers (e.g. Nick Bostrom's *Super-Intelligence*). Second, they are actively engaged in building our future: entrepreneurs like Elon Musk and Peter Thiel are "intentionally tackling fundamental questions about the nature of consciousness and what constitutes the good life, questions that once lived mainly in philosophy departments" (Sathian 2016). Philosophy is moving on with or without the (academic) philosophers. So what should academics do? Straddle the divide: one foot planted on the stable terrain of the university, where free speech and slow thinking are luxuries that are still in force, the other standing in the more transitory networks of the modern-day Republic of Letters where ideas are being field tested. For the philosopher need not only be the Owl of Minerva hooting about what's already been done. The philosopher can also be the robin at work in the light of dawn.

4. FIELD PHILOSOPHY IN A NUTSHELL

The rhizome has no beginning or end; it is always in the middle, between things, interbeing, intermezzo.

Deleuze and Guattari, *A Thousand Plateaus*

Society increasingly consists of networks and interstices, held together by connections that are intermittent, transitory, and subterranean.[2] Communication has become rhizomatic: links lead in all directions, prompting unanticipated exchanges, conflicts, and synergies. Thinking that takes place on the web stands in stark contrast to the linear ideal of argumentation long viewed as 'rigorous' by philosophers. There has always been a risible element here: Virginia Woolf's *To the Lighthouse* mocks Mr Ramsey, an earnest twentieth-century philosopher who struggles to get from L to M and then on to N and perhaps – someday – all the way to R. The information age makes absurd what Woolf already saw as antiquated: disciplinarity, aka the idea of keeping thinking within bounds.

When Deleuze and Guattari speak of rhizomatic politics they are emphasizing a subterranean approach to problems. Philosophy has always had its own subterranean quality. Straightforward challenges to the status quo lead to reaction; often it is better to let insights insinuate themselves. Much of the interaction that field philosophers have with other disciplines and society at large – the work of revealing concealed premises, drawing out implicit contradictions, and connecting disparate insights – remains half-hidden and interstitial in nature. Field philosophy emphasizes the messiness and open-endedness of philosophic work, where thinking often gropes in the dark, and is ready to turn this way or that as we encounter unexpected obstacles and opportunities.

As our title, *Socrates Tenured*, indicates, we think of field philosophers as being housed in the university (thus afforded the free speech protections of tenure), but doing much of their thinking (like Socrates) with people out and about in the world who are struggling to define and solve problems. This distinguishes the field philosopher both from the disciplinary philosopher and what we call the philosopher bureaucrat. The disciplinary philosopher speaks only to fellow academics, usually fellow philosophers. The philosopher bureaucrat has a philosophical education but has left the academy to permanently work in the public or private sector.

Table 6.1 is a simple schematic for what we see as a twenty-first-century ecosystem of philosophy. It is an ecosystem with three distinct species of philosophers, which makes it more robust and resilient than the current near-exclusive monoculture of disciplinary philosophy.

Table 6.1

	Disciplinary philosopher	Field philosopher	Philosopher bureaucrat
Institutional home	Philosophy department	Philosophy department and scattered across the university	Public and private sectors
Primary audience	Academics, especially philosophers	Both academic and non-academic audiences	Predominantly non-academics

With this broader context in mind, we offer the following six characteristics as definitive of field philosophy:

- Goal: help excavate, articulate, discuss, and assess the philosophical dimensions of real-world policy problems.
- Approach: pursue case-based research at the meso-level that begins with problems as defined and contested by the stakeholders involved.
- Audience: the primary audience consists of non-disciplinary stakeholders faced with a live problem. Knowledge is produced in the context of use.
- Method: rather than a method, we speak of rules of thumb, a pluralistic and context-sensitive approach with a bottom-up orientation.
- Evaluation: context-sensitive standards for rigour, and non-disciplinary metrics for assessing success, which in the first instance is defined by one's audience.
- Institutional placement: field philosophy resides on the margins of existing institutions, shuttling between the academy and the larger world; but also seeks to institutionalize itself both within academia and different communities of practice.

In sum: field philosophy begins with problems as defined by non-philosophic actors in real-world settings and seeks to make contributions deemed successful according to more-than-disciplinary standards.

Goals: Field philosophy remains as open as possible to seeing 'the problem' from multiple perspectives and at multiple depths. There is no determination in advance concerning problem definition or the correct framing of philosophical analysis; there is only the assumption that philosophical perspectives are lurking in one place or another. What exactly constitutes 'the right thing to do' is not likely to be evident at first, and is not to be predefined through the application of a theoretical framework or methodology. Fieldwork is a prolonged and communal questioning about what constitutes the goal in each case.

Field philosophers seek to be honest brokers, giving equal consideration to all sides of an issue. But this does not mean that they must limit themselves to a role of neutral observer or commentator or facilitator of dialogue. The field philosopher can also advocate for policies that he or she believes promise the best path to securing the common goods at stake. This requires judgement, which must remain ongoing in case changing conditions warrant a change in direction. In advocating for particular policies, the field philosopher may work closely with stakeholders, including activists. The bonds of trust and obligation formed here are important, but they must not stunt the ongoing exercise of critical thinking or hinder the ultimate goal of serving a common good.

Approach: Initially, philosophers play a marginal role within a given situation. Non-philosophers first define what counts as the problem and the range of acceptable solutions. The philosophical dimensions of the problem will be shot through with legal, historical, political, scientific, technological, cultural, economic, and other dimensions. The learning curve may be steep, requiring research in unaccustomed areas and ongoing interactions with stakeholders. Listening and being actively involved are key activities for earning credibility among the target audiences. Often field philosophers will not have a ready-made platform to gain the ear of their audiences, so they will need to earn their street cred.

By 'meso-level' we mean field philosophy operates on the organizational plane, where general points or policy issues are at stake. This is different from the ancient idea of the philosopher king, where philosophers seek to influence a politician or ruler. But it is also different from the teacher or counselor who focuses on the micro-politics of life, and the disciplinary philosopher who assumes that insights somehow will disseminate. Meso-level work operates at the project level, lasting some months or years, and often involves being embedded in a bureaucratic culture for a period of time (Frodeman 2007). To give some examples of meso-level field work, we have worked with the US Geological Survey, participated in policies pertaining to fracking regulation at the municipal level, helped stakeholder groups in Southwest Colorado address acid mine drainage, and interrogated peer-review practices at federal research funding agencies.

Audiences: The primary audience for the work of field philosophy consists of one or another group of non-philosophers. Field philosophy rejects the linear model of disciplinary knowledge production according to which knowledge is produced in isolation from the context of use and then deposited in a reservoir of peer-reviewed publications for potential users to draw from (Pielke and Byerly 1998). Rather, field philosophers engage in the co-production of knowledge with non-philosophers, where the context of use defines the work to be done (Gibbons et al. 2004). Though disciplinary peers retain a crucial role in field philosophy, the field philosopher recognizes a greater obligation to his or her non-disciplinary peers and the issues they are confronting. Disseminating results to one's disciplinary peers is part of the remit of the field philosopher, especially as a way to improve practices; but it is not sufficient to demonstrate rigour or success.

Method: Field philosophy constitutes a method only in an extended sense: it uses whatever means necessary to accomplish the tasks presented from within the case study. We've found ourselves doing surveys, studying mineral appraisal data, conducting rough cost-benefit assessments, and much more. This might be viewed as opportunistic, and as lacking the rigour of disciplinary research; but we think of it as harmonious with an ameliorative approach, as well as consistent with the bottom-up orientation of casuistry.

Method, after all, presupposes the kinds of disciplinary demarcations that are now passé.

Evaluation: Questions of evaluation are closely linked to questions of goals – to know if something was successful we need to know what its purpose was. To restate the goal: field philosophers embrace an alternative notion of success that places greater value on helping non-philosophers reimagine and work through their own problems. When it comes to fieldwork, there is no preset understanding of what counts as an acceptably rigorous approach. Appropriate rigour will be defined on a case-by-case basis, by economic, temporal, and political exigencies. And it will be defined primarily in terms of how effective the field philosopher is in achieving non-disciplinary goals. Traditional bibliometrics can be used, but they will be less important than alternative, often case-specific means of evaluating impact.

Institutional Placement: Here field philosophy must be at its most recursive. Field philosophic research should include reflection on its own institutional status to ensure that appropriate standards and procedures are in place to assess the non-disciplinary practices. Other philosophy faculty, department chairs, and university administrators, along with a wide variety of actors in the public and private spheres, all become potential stakeholders in a process of building new, often ad hoc institutional frameworks, what Guston (2000) calls boundary organizations. The institution-building aspect of field philosophy also has pedagogical elements, as it must foster a future community of practitioners with some of the unorthodox skills required for successful fieldwork.

5. ALLIES

We are not the only academic philosophers seeking to break with – or to add something onto – the disciplinary model of philosophy. There is an ill-defined contingent of public, rather than applied, philosophers who are experimenting with new, non-disciplinary modes of research. In this section, we briefly survey these philosophical allies, outlining the broad contours of an emergent community. Our message here is twofold. First, these thinkers are doing innovative work that often breaks with the disciplinary mode of research. We have learned a great deal from the transdisciplinary and interdisciplinary activities they're engaged in. But second, this nascent group has not been systematically theorized as a non-disciplinary model of philosophy. Much of this non-disciplinary work remains at the fringes; by not being framed as an alternative model it is in danger of being dismissed as a series of one-offs.

Our argument began (in chapter 1) with what we called Mode 2 philosophy – any approach to philosophical research and teaching that breaks with the disciplinary model of knowledge production and transfer. Mode 2 philosophy is conducted by academic philosophers (or academically trained philosophers) who focus on non-disciplinary audiences, thus distinguishing it from applied philosophy. It strives to fill a social need for greater reflection, a need because of either specialization and fragmentation, unspoken assumptions, or simply a lack of time to reflect.

Allowing for overlap, Mode 2 philosophy may be divided into four main branches:

- *Popular* approaches: these include philosophical cafes, where topics of perennial philosophical interest or contemporary controversial issues are discussed in structured but informal social venues or on the web on blogs, podcasts, etc. Philosophy in the Public Interest, directed by Andrea Houchard at Northern Arizona University, has a range of such programmes, including: Philosophy in the Schools, Environmental Ethics Outreach, Moral Courage (curriculum for at-risk high school students), a Philosophy and Film series, and a Hot Topics Café. Other examples include podcasts and blogs on such sites as Philosophy Now, The Partially Examined Life, and Philosophy Bites.[3]
- *Pedagogical* approaches: this can be divided into two stems: philosophers teaching philosophy outside of the college classroom (e.g., in primary and secondary schools or in prisons) and philosophers incorporating public engagement into their college classes (e.g. service learning and civic engagement). A leading organization here is PLATO (Philosophy Learning and Teaching Organization), which supports introducing philosophy to K–12 students. There are also journals devoted to these issues, including *Childhood and Philosophy* and *Thinking: The Journal of Philosophy for Children*.
- *Interdisciplinary* approaches: these focus on collaboration within the academy. For example, experimental philosophers use the methods of other disciplines (especially cognitive science and sociology) as a fresh way to tackle philosophical questions (Knobe and Nichols 2008). Other philosophers help scientists and engineers think about the philosophical dimensions of their research. The Toolbox Project, led by Michael O'Rourke at Michigan State University and Stephen Crowley at Boise State University, helps interdisciplinary scientific and engineering research teams via a structured survey and discussion designed to elicit different worldviews – with the ultimate goal of fostering improved interdisciplinary collaboration and thus more successful research projects (O'Rourke and Crowley 2012). A similar programme is STIR (Socio-Technical Integration Research). Led by Erik Fisher at Arizona State University, STIR embeds philosophers and other humanists in science and engineering labs in an effort to foster greater

awareness about the social and ethical dimensions of their research so as to conduct more responsible innovation (Fisher and Mahajan 2006).

• *Transdisciplinary* approaches: these often have interdisciplinary elements but treat stakeholders outside of the academy – often policymakers or people engaged in policy discussions – as their primary audience. Paul Thompson at Michigan State University has had a successful career practising what he calls "occasional philosophy" in both the public and private sectors of agriculture (Thompson 2010). Much of our work in field philosophy is also transdisciplinary in orientation.

There are other ways of dividing up approaches to Mode 2 philosophy. For example, projects can stem from a particular grant. They can be more or less methodical with one or another type of protocol, survey instrument, etc. They can result from an invitation by a non-philosophical audience (i.e. demand or supply driven), with projects ranging from explicit, grant-based arrangements for a philosopher to engage with members of a science lab to quite informal social networking and commentary by a philosopher working on a policy issue. And Mode 2 projects can either limit themselves to asking questions and facilitating dialogue or entail offering advice, even advocating for certain actions or outcomes.

With these distinctions in mind, we now turn to some of the main groups promoting Mode 2 philosophy. In the interest of brevity, we'll touch on just three organizations that we think are doing some of the best work to foster communities of Mode 2 practices.

• The Public Philosophy Network (PPN)

PPN grew out of a 2010 workshop hosted by the APA Committee on Public Philosophy (then chaired by Elizabeth Minnich) as part of the Pacific Division APA conference in San Francisco (see Meagher and Feder 2010; Meagher 2013). We were in attendance along with forty other philosophers. The 'catalyst speakers' for the workshop were Andrew Light, John Lachs, Linda Martín Alcoff, Noëlle McAfee, Eduardo Mendieta, William Sullivan, and Nancy Tuana. The workshop had the goals of exploring the meaning and value of public philosophy and building a network of like-minded scholars. At the end, participants advocated three positions: philosophy is a public good and should be practised in and with various publics; public philosophy has the explicit aim of benefiting public life; and public philosophy should be liberatory (i.e. assisting those most vulnerable, particularly through critical analyses of power structures). PPN subsequently formed as an online social network and has as of this writing hosted three conferences of its own: 2011

(Washington, DC), 2013 (Atlanta), and 2015 (San Francisco). The online site serves as a portal to connect over 30 affinity groups and 900 individual members. Although not officially connected with the PPN, the *Public Philosophy Journal* partially grew out of conversations held at PPN conferences.

• Socially Relevant Philosophy of/in Science and Engineering (SRPoiSE)

Also in 2010, the same year as the workshop that birthed the PPN, we travelled to Michigan State University (MSU) for a workshop that explored the idea of a consortium of socially engaged philosophers and philosophy departments. That year, Kathryn Plaisance (of the Centre for Knowledge Integration at the University of Waterloo) and Carla Fehr (Department of Philosophy and Religion Studies at Iowa State University) published a special issue in *Synthese* titled 'Making Philosophy of Science More Socially Relevant'. In it, they introduced the acronym SRPOS (socially relevant philosophy of science), writing "The impetus of this project was a keen sense of missed opportunities for philosophy of science to have a broader social impact." (Fehr and Plaisance 2010).

Through a series of informal social networking and workshops over the next couple of years, SRPoiSE was born. Its core institutional membership consisted initially of MSU, the Rock Ethics Institute at Penn State, the Center for Values in Medicine, Science, and Technology at the University of Texas at Dallas, the Reilly Center for Science, Technology, and Values at the University of Notre Dame, and the Center for Science, Ethics, and Public Policy at the University of Delaware. As of this writing, there are roughly thirty individual members. The mission of SRPoiSE is to "improve the capacity of philosophers of all specializations to collaborate and engage with scientists, engineers, policy-makers, and a wide range of publics to foster epistemically and ethically responsible scientific and technological research."

• The Society for Philosophy of Science in Practice (SPSP)

SPSP was founded in 2005 at the Vancouver meeting of the PSA. More European in orientation than either PPN or SRPoiSE, the founders of SPSP are Henk De Regt (Free University of Amsterdam), Marcel Boumans (University of Amsterdam), Mieke Boon (University of Twente), Hasok Chang (University of Cambridge), and Rachel Ankeny (University of Adelaide). SPSP holds meetings once every year or two and as of 2015 had over 750 members on its mailing list. SPSP can be seen as part of a long evolution. As Reisch (2005) notes, the original impulse behind the *Wiener Kreis* was sociopolitical as well as epistemological in nature; but by the 1950s mainline philosophy of science had become strongly internalist in orientation. One effect of the failure of mid-century philosophy of science to take the larger cultural effects of technoscience seriously was the development of science and technology studies in the 1960s. Thomas Kuhn's 1962 *Structure* marked the beginning of the long slow march of the philosophy of science away from its internalist focus and towards taking history and culture seriously. The founding of SPSP can thus be seen as the next logical step in this process – a response to the deficiencies of mainline twentieth-century philosophy of science, by emphasizing questions attendant to the actual practice of science in the real world.

As with PPN and SRPoiSE, the discussions at SPSP often struggle to break the bonds of disciplinary capture. At the 2015 SPSP meeting in Aarhus, Denmark, for example, there were few reports on what use non-philosophy audiences would take from the analyses being offered, or of how the role of philosopher or the standards for quality philosophic work needed to change in response to these different circumstances. Although these groups are the furthest along in thinking beyond disciplinarity, there is still conceptual work to be done in terms of how to reposition the institutional expressions of philosophy. We think that our notion of field philosophy can help.

6. THE INSTITUTIONS OF ACADEMIC PHILOSOPHY

Philosophy, as the thought of the world, does not appear until reality has completed its formative process, and made itself ready. ... Only in the maturity of reality does the ideal appear as counterpart to the real, apprehends the real world in its substance, and shapes it into an intellectual kingdom.

Hegel

Granting widespread conformity to disciplinary norms, there are still a number of experiments within professional philosophy in what could be called de-disciplinary philosophy. What's missing is the raising of these experiments to institutional awareness and commitment. Until this occurs the creative work being done will bloom and die and bloom again with little built upon past work.

We see scant evidence of this institutional work occurring so far. A survey of the professional organizations of philosophy shows little philosophical thinking about the institutional dimensions of philosophy. The APA runs the three division meetings, publishes job listings, and convenes working groups on middle-level (and certainly important) questions such as the status of women in the profession. In the case of the Philosophy Documentation Center, statistics are collected every couple of years, and the organization manages the back end of various journals and conferences. But one finds no tracking of trends, for example, in whether or to what degree philosophy faculty is attracting federal grants, changing tenure and promotion standards, engaging in applied research, or publishing outside of philosophy journals – questions that we asked in our own 2010 survey (Frodeman 2012). One searches in vain for a philosophical and sociological analysis of the discipline, except as high theory (e.g. Collins 1998). One finds no sustained analysis of whether philosophic academic culture is maladapted to twenty-first-century society. There is not even anything along the lines of the MLA's own societal analyses,[4] as conventional as these have been.

One sign of the profession's current interpretation of its own mission can be found in the position of the APA director. The background of the current (2016) executive director – a bachelor's degree in women's studies and a master's degree in public policy and administration – reflects a creditable openness to different skillsets, as well as the awareness of the need for practical competences in areas such as diversity, communication and development, and non-profit management. This is all to the good. But there are also theoretical dimensions to the problems facing the discipline, which call for historical and sociological analyses of the state of the profession. Institutional organs like the APA should be providing leadership in terms of identifying trends and new opportunities for twenty-first-century philosophy. Bringing in the perspectives of women's studies and public policy represents a potential shift toward a more interdisciplinary view of philosophy – if, indeed, these perspectives issue in new initiatives and approaches at the APA.

There *are* innovations to point to within the institutional status quo. Perhaps the most noteworthy of these are occurring at Arizona State University (ASU). At ASU the process has been largely driven by President Michael Crow. Crow's focus has been on rethinking the function and structure of the research university in the twenty-first century – a point that he has made clear in a number of publications, including *Designing the New American University* (Crow and Dabar 2015). Under his influence both the department and the place of philosophy at ASU have been made over.

Accounts suggest that the process has not gone entirely smoothly. On one occasion, Crow told a gathering at ASU that he considers philosophy to be his biggest failure since his arrival.[5] Different sources describe versions of the following chain of events. Crow visited the philosophy department early in his tenure; since every great university needs a great philosophy programme, what resources were needed to accomplish this? The philosophy faculty responded that they already had a first class programme. Crow argued that ASU needed a philosophy programme more responsive to community needs. The faculty again demurred, and in response Crow placed philosophy within a new unit, the School of Historical, Philosophical and Religious Studies – on some accounts consigning the programme to a back water.[6] Some of the philosophers who remained committed to a standard approach to philosophy found employment at other institutions. In 2015, there were some ten full-time employees (FTEs) in philosophy. Its PhD programme was inaugurated in 2000, but in 2009, at the time of the Great Recession, philosophy was among the programmes affected, and the PhD programme stopped admitting students. In 2013, the programme was restarted, but was now reimagined in terms of a "newly redesigned Ph.D. in Philosophy [that] features a focus on Practical and Applied Philosophy."

This, however, constitutes only part of the philosophical activities at ASU. There are a number of philosophers embedded within other schools across campus, in units such as the School of Earth and Space Exploration (SESE) and the School of Life Sciences (SOLS). SOLS includes the philosophers Ben Minteer, Jane Maienschein, and Richard Creath, all of whom are involved in interdisciplinary and transdisciplinary work at the project level. Minteer works with biologists, wildlife managers, and animal advocates on questions of extinction, loss, recovery, and resurrection. And Maienschein runs a lab that combines "analysis of the epistemological standards, theories, laboratory practices and experimental approaches with study of the people, institutions, and changing social, political, and legal context in which science thrives" (see her website).

Michigan State offers a somewhat similar profile. It has a stronger philosophy department than ASU – double the number of philosophers (around twenty) – and compared with ASU there are fewer philosophers distributed across campus. But one finds philosophers embedded within residential colleges like Lyman Briggs, which focuses on a science curriculum. Like ASU, MSU philosophers regularly apply for NSF grants and have a number of ongoing interdisciplinary projects. Paul Thompson, as we have noted before, has extensive experience working on questions in agricultural ethics with both the public and private sectors. Michael O'Rourke and colleagues have run 'toolbox' workshops that help interdisciplinary teams become more aware of their own epistemological and metaphysical assumptions. On the other hand, from the accounts we have gathered there has been little attempt to view these two roles, the disciplinary and the de-disciplinary, as complementary to one another, or that there should be circulation between the two types of activity. Nor does either university make a self-conscious effort to train PhDs in philosophy for work in the public or private sectors.

Perhaps the most noteworthy omission is this: while Crow is notably self-conscious in both speech and in print about describing ASU as a new model for the twenty-first-century research university, one searches in vain for published books or articles on the ASU (or MSU) model of philosophy or the humanities. Philosophers here and elsewhere may be distributed across a number of schools, but they do not meet in any regular or organized fashion, and are not treated as a unit with a common mission of doing something like de-disciplinary philosophy. There is no organized, theoretically conceived connection between philosophers in the various schools.

7. THE LIBERTARIAN PREJUDICES OF PHILOSOPHY

We've argued that philosophy within the modern research university has honoured Socrates more in word than in deed; but there is one element of Socrates' legacy to which philosophers have remained true: we go our own way. Autonomy is part of our DNA. In terms of our professional instincts if not political commitments we are libertarians. The department augments this, functioning as a machine for individualization, social isolation, and a certain kind of epistemic closure.

We should acknowledge the philosophical reasons for this unphilosophical attitude towards the institution. In part it reflects the academic's disdain for groupthink and a jealous sense of one's autonomy. The structure of the department, what Graff called the field coverage model, ensures the maximum amount of non-interference and self-determination: no one is going to tell *me* what to teach or research. This structure also reflects the historical, analytic, and individualistic bias of philosophers. Descartes placed the individual cognizer as prior to social structures. The same attitude pervades political philosophy: social contract theory posits Robinson Crusoe as existing prior to the establishment of any state or social structure – childhood, language, and culture notwithstanding.

Stay with political philosophy for a moment, and compare John Rawls's book *A Theory of Justice* with Daniel Callahan's essay 'Bioethics as a Discipline'. Both were published in 1971. The two offer a fork in the organizational history of philosophy.[7] Concerning Rawls, Kuklick notes, "Just as Quine had eschewed linguistics when he spoke of translation, so Rawls ignored the collective experience of attempts to construct a just society. In real life, ignorance of one's actual cultural locus – one's sex, age, social status, and race – would disqualify one from politics; Rawls made such ignorance the *sine qua non* of acceptable participation" (pp. 263–64). The social isolation of disciplinarity shaped the content of philosophical thought: disciplinary purity prepared the ground for the abstractions of Rawls's thought. Callahan, on the other hand, embraced the mangle of practice and the muddiness of what he called "the realities of life." He recognized how working in real time with nonphilosophers fundamentally altered the shape and content of his philosophical work. These experiences forced him to confront the "disciplinary reductionism" unconsciously practised by philosophers – the habit of isolating something recognizably philosophical in the hurly-burly of a real-world issue, distilling it out, labelling it *the issue*, and speaking authoritatively about it.

This habit produced knowledge that non-philosophers often find useless, because there was no purely philosophical space in the problems they confronted where it could be slotted in. This prompted Callahan to argue that the bioethicists must practice "a different kind of rigor" when working in transdisciplinary and interdisciplinary contexts (Callahan 1973). Knowledge production and cognitive activity must be organized differently when one leaves the department. While Rawls built castles in the sky, Callahan laid out a programme for training new kinds of philosophers and institutionalizing a new kind of philosophical practice.

Nietzsche already noted in *Beyond Good and Evil* (1886) that the scope of knowledge had expanded to such an extent that the philosopher grows weary in learning and "allows himself to be detained somewhere to become a 'specialist' – so he never attains his proper level." But in addition to its epistemic element, disciplinarity is also rooted in personal psychology. To wander beyond its confines is to invite disorientation and vulnerability. Nonetheless, we can accommodate ourselves to not being authoritative. There is another way to philosophize – that is interstitial, horizontal, and reciprocal. Field philosophy breaks with the model of the sage producing a book or article to be delivered downstream to consumers. Instead we can become partners, working as a member of a team, seeking out insertion points, and adapting our contributions to what is learned along the way.

NOTES

1. Although ten years ago we did assist in the creation of a field station in philosophy in southern Chile: http://chile.unt.edu/our-approach/pan-american-environmental-philosophy-network.

2. "Parts of this argument appeared in Frodeman, Robert, Briggle, Adam, Holbrook, J. Britt, 2012. 'Philosophy in the Age of Neoliberalism'. *Social Epistemology*, vol. 26, issue 3&4.

3. A good list of podcasts can be found here: http://people.wku.edu/michael.seidler/50podcasts.pdf.

4. See, for instance, the 2014 Report of the Task Force on Doctoral Study in Modern Language and Literature, at https://www.mla.org/Resources/Research/Surveys-Reports-and-Other-Documents/Staffing-Salaries-and-Other-Professional-Issues/Report-of-the-Task-Force-on-Doctoral-Study-in-Modern-Language-and-Literature-2014.

5. For understandable reasons our sources have asked to remain anonymous.

6. The majority, but not all of our sources agreed with this account.

7. We might note another road, stemming from Dewey, that gave rise to the policy sciences, a non-positivistic approach to solving social problems with normative dimensions. The policy sciences can be seen as a kind of engaged philosophy by another name.

Chapter 7

The Philosophy of Impact

The previous chapter described our approach to increasing the impact of philosophy on society, what we call field philosophy. That chapter also opened up a larger set of questions concerning societal demands that research be more accountable to society. In other words, a reversal is in order: questions of the impact of philosophy imply the development of a philosophy of accountability and impact. What is meant by 'impact'? What are the different types of impact? Can we rank impacts as more or less good or bad? And is there a case to be made for knowledge that does not have an impact?

'Impact' is a dubious choice of metaphor for those seeking greater accountability in research (or for that matter, education). The term is too Newtonian, suggesting the effects of a car crash, when most outcomes are much more indirect and varied than that. Terms like 'influence' or 'sway' better represent the complex processes involved. We cannot measure the impact of research in the same way we can measure the impact of a car hitting a wall; there is no equation like $F = ma$ to be had. Moreover, any set of impact metrics contains interpretations and value-laden assumptions. Black-boxing these (epistemic, political, and metaphysical) values in a spurious attempt at being 'scientific' doesn't make them disappear – except perhaps as a matter of political reality (Briggle 2014).

In recognition of such facts, the Leiden Manifesto outlines ten principles for using bibliometrics to evaluate research (Hicks et al. 2015). A code of conduct for bibliometricians is a good start. But a philosophy of impact (by whatever name) needs to consider a wide range of philosophical assumptions lying within contemporary impact debates (Frodeman 2016). Consider the assumptions embedded in requests for broader impacts. Has 'broader impact' largely been defined as economic in nature because we assume that the one thing everyone can agree on is that they would like more money? But aren't we at the end of that gesture, as it becomes clearer that there are trade-offs for every increase in income (e.g., China's struggle in balancing economic growth and environmental health)? Again: consider the educational metric of on-time graduation. Couldn't it be a sign of growing self-knowledge for a student to change majors, and thus delay graduation, as they discover hidden interests or talents? What's a semester or two of tuition and lost salary compared with a chance at a greater degree of happiness and satisfaction at work across a lifetime?

These matters, which strike us as Nietzschean in tenor, constitute the question of pertinent knowledge. What is the role of knowledge in a good life? Is a 'knowledge society' really an unalloyed good? Have we thought through what 'useful' or 'practical' means? Even to raise such questions puts one at risk of being dismissed as a crank, or lumped in with the reactionary and the hidebound. Right-thinking people know that life should consist of continual progress; the ever accelerating pursuit of knowledge is an obvious good. Few seem to have realized that this constitutes a practical programme for human deification (cf. Fuller 2011). Nonetheless, let us ask: Is knowledge (of whatever kind) always pertinent? Or are there times when knowledge production becomes quite impertinent, a distraction, a dodge, or a danger (cf. Shattuck 1997)? Daniel Sarewitz once noted how calls for additional scientific research can serve as an 'out' for politicians not wanting to make hard decisions (Sarewitz 1996). Now the question has taken on more dangerous hues, popping up in unaccustomed places, even among the technological elite. The locus classicus of such fears is Bill Joy's 2000 essay, 'Why the Future Doesn't Need Us', but more recently other techno-luminaries such as Elon Musk, Bill Gates, and Nick Bostrom have all voiced concerns about the dangers of one particular type of technological progress – artificial intelligence.

We cannot do more than broach such questions here. But this, in fact, is our main point: philosophy needs to ask these questions in as many venues as possible, both to promote its own relevance and vitality, and to challenge a society that too easily takes 'impact' and 'innovation' as unimpeachable goods. This topic presents philosophy with an opportunity for theoretical work, exploring an intellectual space that draws from elements of social and political philosophy, epistemology, the philosophy of interdisciplinarity, the philosophy of action, and science policy.

In what follows we first survey the recent history of thinking on impact, a tradition which finds its home in the policy sciences and science policy. We then review the history of discussions of impact in philosophy and the humanities. We follow this with a suggestion for a middle way in impact, one that centres on intentions and processes rather than products and results. We call our approach, the act of expanding our moral imagination, curiosity *plus respicere*.

ZEN AND THE ART OF KNOWLEDGE PRODUCTION

What good is the pursuit of knowledge? The question is largely personal when the investigator is self-supporting. But the question becomes social when someone else supports the research: what kind of return will they see on their investment? It turns political when public funds are at stake: does this research serve the common good? And it becomes existential when research can put an end to civilization, as became possible with the advent of the atom bomb at the end of World War II. This last point is one that has gained renewed pertinence with recent developments in genetics, nanotechnology, and artificial intelligence.

In his 1945 report, *Science – the Endless Frontier*, Vannevar Bush gave the first systematic apology for publicly funded research – one that gave a deceptively powerful answer to the question of impact. Bush made the case for institutionalizing the public funding of research. He wanted to keep public funding flowing in the aftermath of the war. But he also sought to remove the constraints that had kept researchers bound to projects geared towards immediate applications, such as improved radar and bombs. His pamphlet would become the blueprint for the US NSF. Bush offers a distinctive argument. He claims, "Basic research is performed without thought of practical ends." Yet without basic research, the reservoir of knowledge necessary for improvements in health, security, education, and communication will dry up. "Statistically it is certain that important and highly useful discoveries will result from some fraction of the undertakings in basic science; but the results of any one particular investigation cannot be predicted with accuracy."

The argument reads like a koan. Research is impractically practical. Knowledge is pursued for its intrinsic value – while yielding practical results. Impacts are never intended, but they always come – just as the warbler appears, unbidden, after each winter. Science can predict everything except its own results. It is the eye that sees everything but itself. This peculiar argument became the default contract between science and society. It created a system of accountability premised on disciplinary standards. Accountability was defined by peer review: scientific research was judged by one's disciplinary peers. If those peers found this research to be good by their standards, then the research was ipso facto good for society. Peer review defined the extent of the obligations of the scientist: Just do good research – there is no further need to think about its broader impacts. Society is free to draw from the results as it sees fit, turning research outputs into impacts. How that unfolds is not the business of the researcher.

The pathway to impact was thus shrouded. Moreover, the path must remain dark; to light the path was to eliminate the shadows that success was

predicated upon. Those without faith need only to look at the track record
of success. And since outputs automatically (though unpredictably) yielded
impacts, there was no need to track or measure impacts. It was, rather, a calculated investment in serendipity. In Bush's terms, it is "the free play of free
intellects, working on subjects of their own choice, in the manner dictated by
their curiosity for exploration of the unknown," that in a happy coincidence
yields a healthier, wealthier society. In terms of accounting we only needed
to tally peer-reviewed articles. High quality (i.e. disciplinary, peer-reviewed)
research was both the necessary and sufficient condition for highly impactful
research.

This is the hydraulic answer to the question of impact. Knowledge flows
from the university down a pressure gradient to society. Michael Polanyi
(1963) would later put the point in market terms. Research works through
the "spontaneous coordination of independent initiatives." You can kill science, Polanyi says, but you cannot shape or direct it, because it "can advance
only by essentially unpredictable steps, pursuing problems of its own, and
the practical benefits of these advances will be incidental and hence doubly
unpredictable." This conclusion might appear "adrift, irresponsible, selfish ...
chaotic" to those "socialists" who want to plan the progress of society. But
freedom and its attendant unpredictability are necessary, even though that
makes them the "less attractive aspects of a noble enterprise."

This serendipity model, coupled to the mechanics of water seeking its own
level, has served as our default philosophy of impact. Its Zen-like quality has
resisted analysis. Just as the plant 'knows' how to blossom and we 'know'
how to breathe, this kind of knowing cannot be set down in propositions. The
knowing is in the doing. We know how research impacts society, because it
does.

From a philosophical point of view it's an intriguing answer. There's much
to like here: it shows respect for the wellsprings of creativity, as well as disdain for the often obtuse nature of political interference. But from the point
of view of public accountability it has become inadequate. Strip away its Zen-
like qualities and it reads as 'trust us' – an unacceptable answer in sceptical
times. In a post-Cold War age of budget cuts and increasing accountability,
funding institutions need a more persuasive account of impact. They need
to put what Bush and Polanyi might simply have called "the way" into for-
mulae, methods, and metrics. They have to model it so that impacts can be
brought about more efficiently. And they want a more guaranteed return on
investment.

The defining feature of the accountability culture is the quest for an
explanation of impact that does not rely on serendipity. To 'give an account'
means to spell out how impacts happen, not to just wave your hands over
some black-boxed alchemical transformation. Never mind the irony that the

accountability culture that Polanyi identified as "socialist" is today driven by a neoliberal agenda. Or that it displays an enduring faith in the market, operating by its own type of serendipity, that of the invisible hand. We see two extreme positions here, neither of which is workable. At one end, researchers can dig their heels in and insist on the mysterious workings of serendipity. This is no longer politically viable. At the other end, funding agencies (private or public) can demand a complete (usually econometric) reckoning of every pathway and every impact. This is not epistemologically viable. For Bush and Polanyi are right that, at some level, the workings of research and its complex interactions with society are simply unknowable, and attempts to define or control these interactions too tightly will backfire.

THE HUMANITIES: FROM VALUE TO IMPACT

Demands for accountability have always been part of the rhetoric of public funding. But in the United States, social pressures for the demonstration of impact increased markedly in the mid-1990s. One of this shift was the introduction of the Broader Impacts Criterion at NSF in 1997, where peer reviewers were now asked to give equal weight to both the intellectual merit and the broader impacts of research proposals. A second was the creation of the SciSIP programme at NSF in 2006 – an attempt at having the eye see itself. SciSIP supports research that creates new "models, analytical tools, data and metrics that can be applied in the science policy decision making process." Its goal is "to predict the likely returns from future R&D investments" so that more "fruitful policies" can be made (SciSIP 2015). A third marker, from outside the United States, occurred in the 2010 creation of the United Kingdom's Research Excellence Framework out of the previous Research Assessment Exercise, which itself dated back to 1986. Upon its completion in 2014, 154 universities submitted bibliometric data and more than 6,900 case studies on the impact of their research, in areas ranging from physics to the classics.

In Europe and the United Kingdom, the impact agenda applies to both the sciences and the humanities. One example of the latter is the Humanities in the European Research Area (HERA), which consists of a partnership of twenty-one European Humanities Research Councils and the European Commission. In contrast with the United States' National Endowment for the Humanities, HERA projects commonly work with the STEM disciplines. They also share the latter's emphasis on impact. Moreover, the European equivalents of the NSF, Horizon 2020 and the European Research Area, explicitly include the social sciences and humanities: as noted on the website of the former, "Social Sciences and Humanities (SSH) research is fully integrated into each of the general objectives of Horizon 2020." This has led to the creation a literature on humanities impact largely absent from the United States (e.g. Benneworth 2015).

In the United States, the impact agenda has largely been confined to the STEM disciplines – the National Endowment of the Humanities and the National Endowment of the Arts, constantly threatened with elimination by Congress, steer clear of anything controversial. But even in the STEM disciplines the conversation is much less developed than in Europe. This is in part an artefact of American federalism, where power is devolved from Washington to the fifty states. Such a system makes universal mandates difficult to achieve. But whether at the state or federal level, it's inevitable that philosophy and the humanities will eventually be called to account. We have found that American humanists are ignorant of the fact that in the United Kingdom the audit culture already includes the humanities. The few American humanists who have been paying attention greet the possibility of an impact regime with a mixture of dread and condescension – the imposition of half-brained, narrow, impatient, neoliberal metrics onto fields of inquiry that cannot be reduced to mere "utility" (cf. Harpham 2011a).

We understand this reaction. But scorn won't make these demands go away. And while these demands are discomforting, the impact regime could also contribute to a renaissance for American humanities. Lemonade can be made from these lemons. And it should be: for the current system of humanities research is far from healthy. As we noted earlier, by some estimates 85% of research in the humanities goes uncited and perhaps even unread by anyone. The relationship between the production and the consumption of humanities research is largely unthought of and could use a jolt.

Humanists do a poor job of accounting for the impact of the humanities. They are too busy producing scholarship to consider its social worth, and they do not consider meta-analyses of the effects of that work to be part of their remit. When they do try to explain the virtues of their work they lean on old categories of thought – speaking, for instance about the value (in contrast to the impact) of the humanities, and often sounding defensive and diffident on the subject. The result, as Donoghue (2008) describes it, is "rhetoric that has grown hackneyed and ineffective."

Now, apologies for philosophy go back to, well, the *Apology*. Of course, *apologia* in Latin means a defence; but there has been something 'apologetic' in the modern sense of the term to most defences. Socrates' own apology was quite aggressive, claiming that the state and civil society were badly in need of his efforts, even to the point of demanding that his judicial 'penalty' be a lifetime sinecure. In a series of essays (e.g., Strauss 1952), Leo Strauss emphasized the centrality of the relation between the philosopher and the polis to Plato's thought, to the point that the dialogue form itself serves as a mediator between the two spheres, rendering it safe for philosophy to have a public role.

The modern origin for defences of the humanities is Matthew Arnold's 1869 *Culture and Anarchy*. Arnold is usually interpreted as a defender of the effete joys of high culture (in part because of the unfortunate phrase "sweetness and light"). In fact, his interest in promoting culture was motivated by the social strife engendered by the co-development of democracy and industrialization in Britain. Arnold sought a way to overcome the selfishness and parochialism (on all sides) of class interests represented by the 1866 Hyde Park Riots. Then, Arnold has been used to bolster cultural elitism – the humanities secure in their ivory tower, an exemplar of excellence for the hoi polloi. Thus the words of the 2004 presidential address at the MLA: "We have nothing to offer but the sweetness of reason and the light of learning."

Arnold actually framed the matter as a clash between culture and anarchy. Culture meant what the Germans call *Bildung*, cultivated by consulting the best that has been thought and said; individuals used these ideas to rid themselves of bad habits, just as society used them to solve its problems. Anarchy is his term for a crass libertarianism that worships "freedom in and for itself." The emerging industrial democracies promised freedom for the mass of men, but they were prone to waste this gift. Rather than practising the hard art of self-improvement, they chose to gratify base pleasures. Arnold challenged the usefulness of material utility: we needed the "light to enable us to look beyond machinery to the end for which machinery is valuable." Arnold was an elitist, but it was the elitism of someone who believed it was possible to identify better and worse ways to live, what he called culture, and which we call philosophy and the humanities. It is an idea of the humanities as consisting of something more than disciplinary connoisseurship.

A survey of the contemporary literature offering justifications for philosophy and the humanities includes Nussbaum (2010), Small (2013), Deresiewicz (2015), and Zakaria (2015). Their defences can be divided into:

- The pedagogic: the humanities teach critical thinking, useful in a wide variety of settings. Skill at reading, writing, and speaking with clarity and power never lose its relevance; the humanities concentrate upon and exemplify these skills.
- The intrinsic: scholarship serves the innate values of truth, beauty, and goodness. The humanities have and need no larger purpose or justification, even in terms of the cultivation of the individual.
- The civic: the humanities help us to appreciate the depth, commonality, and diversity of the human experience, enabling us to become both better persons and more capable democratic citizens.

We would add three others to this list:

- The provocative: the humanities criticize tradition and authority and foster critical public dialogue about community ideals. Socrates represents the exemplar, the philosopher as gadfly.
- The Aristotelian: the highest good for human beings is contemplation – attending to the experience of things without seeking to turn them into something else. (On some accounts this is what the intrinsic justification above is really about.)
- The Hegelian: science and technology, long thought to be the opposite of the humanities, have transcended themselves. In a reversal worthy of the Cunning of Reason, technoscience now raises philosophical questions of all types.

This list could be elaborated upon, but any such discussion should not distract us from the key question: how are these values realized or implemented? Far too often when humanists catalogue the worth of their research, they assume that it's packaged and delivered via the disciplinary model of passive diffusion.

The limitations of 'values' talk is highlighted by Ivan Illich's comments about radical monopolies (Illich 1973). There isn't a monopoly of any single automobile brand on the roads. There are Fords, Chevys, Toyotas, etc. – a wide range of choices, one might say. But as a mode of transportation, automobiles exercise a radical monopoly in the sense that they marginalize other modes of conveyance such as walking or bicycling. Auto-centric development is premised on highways, parking lots, big yards, and long distances, creating an inhospitable environment for anyone trying to move around by pedal or foot. We see a similar diversity *and* uniformity of activities and approaches in philosophic research. At one level, there is a riotous diversity; but at another degree of granularity these efforts are (nearly) all of one kind. Monographs and articles operate within the same institutional space, the disciplinary space of professors who treat knowledge as an academic exercise. Work lies behind a wall separating an inside audience of one's peers from other disciplines and society outside. And just as cars tend to shoulder out pedestrians, this disciplinary focus tends to shoulder out alternative modes of research. Working outside of one's peer community is seen as dubious – insufficiently rigorous, mere popularizing, or the dreaded category of service. Or worse, a betrayal of 'real' scholarship.

Unacquainted of Vannevar Bush, philosophers and humanists have nonetheless embraced his trickle-down model of impact. This is where they get in trouble with the culture of accountability: the political class increasingly wants to see researchers demonstrate how their research benefits society, while talk of the 'value' of the humanities relies on the passive diffusion of knowledge, leaving the question unaddressed. "Yes," we imagine a state legislator saying, "you wrote an article about the ethical issues raised by cloning. But how, precisely, does that article do society any good? Don't just tell me what the value is you are aiming at, but walk me down the path to impact where I can see that value *realized*." We don't merely need accounts of the values of the humanities; we need to understand how those values are implemented and mobilized. How, practically speaking, are these values supposed to be realized for people who are not humanities scholars? Without such attentiveness to audience, institution, and practice we risk cataloguing the values of the humanities like we might catalogue the jewels in a treasure box that remains locked at the bottom of the sea. The outside world doesn't just want to know what treasures we have; they want to know how to use them wisely.

Talk of values depends on the great hypothetical: "*If* people were meaning-fully engaged with our research, then they would derive these values (pick from the taxonomy above) from the experience." But are they engaged? Have they been influenced? How do you know? Do you have a plan to get them engaged? The disciplinary model of research does not have a reply to these questions other than the serendipity of trickle-down diffusion. We will be called on to do better than that.

ALTERNATIVE FUTURES FOR THE HUMANITIES

How will the humanistic values listed above fare as the impact regime takes hold in the United States? What strategies might humanists use to chart a path forward?

Teaching Only. One possibility is that the only surviving element will be the pedagogical value of the humanities. Research in the humanities could go largely extinct, as state legislatures decide we don't need the 16,772 article about Shakespeare or Aristotle (cf. Harpham 2011b). US culture is steeped in the idea that science is foundational to social progress. Elected leaders are products of that culture, and so while they might press scientists for more demonstrable impacts, they will remain sympathetic to the importance of scientific research. No such reservoir of sympathy exists for research in the humanities. The situation once was different: in an earlier time conserva-tives as much as liberals had cause to support the humanities. The twentieth century was an era of great ideologies; the humanities served as a bulwark against both communism and fascism. But with the end of the Cold War the humanities became more firmly typecast as liberal in orientation, a condition that imperils them today.

Humanities research will survive at the elite private institutions with large endowments. But a massive scale back is a possibility at public institutions, even ones as prestigious as the University of Wisconsin. Those who make budgets and organize academic life may choose to divorce the pedagogic from the scholarly, saving the former and ditching the latter – the direction that Wisconsin governor Scott Walker is taking his state. The only real value of the humanities, it might be claimed, is in the classroom experience as part of a well-rounded curriculum. It is a far better use of the professorate that they devote their time to teaching, rather than letting them spend half their time producing scholarship that few will ever read.

Intrinsic Value. One reason why the humanities lack the political sympathy that the sciences enjoy is their appeal to the 'intrinsic value' argument. As Benneworth (2015) frames it, the humanities disciplines are "convinced that they are 'useless' or even that utility compromises their essential values." Vannevar Bush may well have viewed science as its own reward, but he didn't pitch it as such. He argued that we should let scientists pursue the projects that they want, but only because this is the means to improved societal health, wealth, and happiness (which is why he spoke of "basic" rather than "pure" research). By contrast, humanists embrace the mentality of *l' art pour l' art*. Serendipity may be very unintentional and indirect, but at least it's a story about impact. Intrinsic value, on the other hand, takes impact off the table.

Intrinsic value accounts of the humanities tend to be simultaneously pompous and muddled. For example, the literary theorist Stanley Fish argued in the *New York Times* that the humanities should be divorced from all that is "useful":

> To the question "of what use are the humanities?" the only honest answer is none whatsoever. And it is an answer that brings honor to its subject. Justification, after all, confers value on an activity from a perspective outside its performance. An activity that cannot be justified is an activity that refuses to regard itself as instrumental to some larger good. The humanities are their own good. (Fish 2008)

What could this possibly mean? That the humanities are their own good in the sense of, what, private amusement? Personal religious ecstasy? With no larger effect on the character of the individuals who study them, or on their students or associates? One could replace the humanities with stamp collection in Fish's statement and get the same amount of sense out of it. By refusing appeal to a notion of a good, Fish cannot distinguish the humanities from the most trivial of hobbies. Part of Fish's problem lies in an assumed dualism between ends and means: the means are nothing but a conduit to some distinct end. But why not think of the humanities–society relationship as a practice where means and ends are all tangled up with one another? In a practice like preparing a meal, the chopping of vegetables and pouring of wine are both intrinsically and instrumentally valuable. The value is in the doing *and* in the end result. A strict separation of means from ends is the kind of thinking humanists need to challenge, not assume.

In the end, Fish's stance comes down to that of an effete, and doubtlessly wealthy, aesthete – intellectually wanting, and more than a little condescending to mere mortals blind to the pleasures of Milton. Today, such an attitude is politically unsustainable. Most institutions of higher learning are publicly owned, and all – through student financial aid and government-sponsored research – are to one degree or another publicly funded. Fish's rhetoric of the "only honest answer" is unlikely to find much sympathy among the citizens and their representatives, who support universities because of their contributions to matters of common concern.

Trickle-Down Humanities. The humanities could chart a future in the age of impact by simply embracing the serendipity model of impact that served the STEM disciplines for so long. All they have to do is transform their unconscious reliance on this model into their explicit impact story. This 'trickle-down' account would be the low-cost option, because it would not require developing alternatives to the disciplinary model of knowledge production and evaluation. To see what such a strategy might look like, we can examine two Bush-like serendipity arguments made by philosophers.

Rudolph Carnap argued that "philosophy leads to an improvement in scientific ways of thinking and thereby to a better understanding of all that is going on in the world, both in nature and society; this understanding in turn serves to improve human life" (1963, pp. 23–24). He adds a prior layer of indirectness before impact: whereas Bush started with science, Carnap starts with philosophy. First, we get the philosophy right, which helps us do better science, which then improves society. The pathway to impact is: philosophy → physics → chemistry → biology → medicine → greater societal happiness and health.

In this framing, philosophy would not just model itself on basic science, but also put itself at the very base of basic science. As Albert Borgmann (1995) notes:

> The modern philosophical project ... emulated and tried to surpass the sciences and the way they contribute to the betterment of society. The leading idea was that involvement in the daily vexations of life is largely futile. Retiring to the laboratory and finding a vaccination for polio is to make a more beneficial contribution to society than to ease the pain and hold the hands of hundreds of polio victims. (p. 298)

The philosopher's armchair is logically prior to the scientist's lab. Before the scientist can make progress in conquering nature for the relief of man's estate, the philosopher must clear the conceptual ground. This requires retiring from the fray of action into the realm of thought. Here 'base', somewhat ironically, is understood in something like Marxist terms: the base of ideas underlies and determines the kind of superstructure of action and things that sits atop it.

Borgmann argues that this trickle-down model is premised on the idea that we can decipher nature's original language and that this is necessary for scientific progress in solving problems like polio. Of course, the quest to mirror nature never panned out: "The search for absolute foundations of what is real and can count as true was misguided and had sequestered philosophy from the conversation of humanity to a barren enclave of professionalism" (Borgmann 1995). Philosophers ran into a quagmire of incommensurable realities, as their discourse devolved into the jargon of the genius contest. Meanwhile, science progressed just fine without philosophers giving it a push from below.

But didn't Bacon and Descartes provide the needed push by establishing the modern scientific picture of the world? Beliefs, worldviews, and even modern science itself come from somewhere. Ideas and practices only change when people come to see themselves in terms of new ideas and practices. New policies are adopted only when new reasons seem compelling. What is the source of this cultural change? Whatever it is, it would be the most impactful thing imaginable, because it would define the shape and substance of thought and action. Baird Callicott (1999) argues that the source is philosophy. Using environmental philosophy as his example, Callicott claims that actions, from personal consumption to public policies, are determined by an underlying worldview, an "ambient intellectual ether" through which we make sense of our experience. Worldviews, in turn, come to life through the aid of philosophy:

> We philosophers are the midwives assisting the birth of new cultural notions and associated norms. In so doing we help to change our culture's worldview and ethos. Therefore, since all human actions are carried out and find their meaning and significance in a cultural ambience of ideas, we speculative environmental philosophers are inescapably environmental activists.

Callicott claims that core elements of the Western worldview – for example, individual rights, atomism, the soul, dualism, and mechanism – are rooted in philosophical thinking from the pre-Socratics onward. Indeed, 'impact' itself is already a philosophical notion – another variation on the mechanical theme of externally related particles transferring kinetic energy back and forth. According to Callicott, philosophers "give voice to the otherwise inchoate and inarticulate thoughts and feelings in our changing cultural Zeitgeist." They may not, then, be the ultimate source of cultural change, but they train the erratic movements of a nameless impulse into the conceptual schema of a new world order, which in turn guides behaviour. John Locke, for example, turned democratic urges into a theory and praxis of public life. Adam Smith articulated and justified an emerging commercial society.

Callicott thinks we need a cure for the modern worldview. We need "a new holistic, non-anthropocentric environmental ethic." Philosophy, with its once innovative view of humans as intrinsically valuable, has done more than anything else to eradicate slavery. And philosophy will eventually eradicate the "wonton destruction of the nonhuman world" because it will help birth a new worldview in which nature has intrinsic value. This will be an impact on the grandest scale, and it will come about through the work of a cadre of intellectual elites filling pages of journals and books with words as Callicott argues:

> My point is that environmental philosophers should not feel compelled to stop thinking, talking, and writing about environmental ethics, and go *do* something about it instead. ... In thinking, talking, and writing about environmental ethics, environmental philosophers already have their shoulders to the wheel, helping to reconfigure the prevailing cultural worldview and thus helping to push general practice in the direction of environmental responsibility.

CURIOSITY PLUS

Callicott's account is a rare reflection on the question of impact. But we suspect that if it were widely adopted by humanists its political fate would be the same as the serendipity story in the STEM disciplines. Indeed, his tale is even harder to swallow than Bush's story. He wants us to believe that when we lift the lid on the enormous dynamo of culture we discover that the engine is little more than a few hamsters on their wheels. Are we supposed to believe that the philosophers' fingers on the keyboard are the ethereal wings of a butterfly conjuring the far away hurricane of a new worldview? What's more, Callicott's argument about the origins of the modern worldview is built using people like Plato and Descartes, thinkers who were working in pre-disciplinary modes of intellectual activity before the modern research university. Are disciplinary scholars working in a knowledge-saturated environment really going to have the same kind of success?

Instead of pushing forward with an impact story grounded in passive diffusion, we propose another model, what can be called 'curiosity *plus respicere*'. This term is inspired by Carl Mitcham's call for engineers to manifest a sense of duty *plus respicere*, the "duty to take more into account" (from the Latin *respicere*, to look again; cf. respect, to treat with regard or esteem). Mitcham (1994) argues that engineers have an obligation not only to consider the narrow functional goals of their work but also the potential broader social, economic, and environmental side effects. For example, don't just evaluate asbestos as an insulator, but also keep an eye out for it being a material with potential health impacts. This is a call to enlarge one's moral imagination. One broadens the scope of intended impacts by taking more factors and more values into account in the upstream and midstream facets of research.

What we have in mind is a model that still allows humanities scholars to follow their curiosity, but to do so in a way that takes more into account. They would be asked to consider the audiences they want to influence, the kinds of change they want to inspire, and the mechanisms or pathways that might bring those changes about. This could take the form similar to the "pathways to impact statement" of the Research Councils of the United Kingdom, built into the research plan outline of a scholar's work. And note how this would also entail the cultivation of a new sense of the scholarly self, with new virtues more like the entrepreneurial intellectual who is adept at speaking to multiple audiences via multiple media. This would also introduce an important change to the serendipity model. Callicott tells the story of Dave Foreman, founder of the radical activist organization *Earth First!*, who once dismissed philosophizing as "cheap talk." But Foreman later listed environmental philosophy as the most important force influencing activists and conservationists. For Callicott, Foreman plays a crucial role as a "thought-to-action-translator" – turning the ideas of philosophers into new policies. Presumably, this role could be played by several kinds of people (not just activists) including lawyers, scientists, teachers, politicians – and philosophers.

The political problem for the serendipity model is that it leaves the relationships between researcher and audience entirely up to chance. It is a *passive* diffusion model because nothing is done proactively by the researcher to facilitate the targeting or transferring of ideas to the idea translators. In a disciplinary regime where so much research languishes unread, or read only by peers, this model offers no means to prevent the fruit of philosophical thought from rotting on the vine simply because no one knew it was there and so no one bothered to pluck it. In contrast, the curiosity *plus respicere* model would have researchers actively seeking to identify 'translators' interested in and influenced by their work. Implementation plans would become an intrinsic part of their research, raising the kind of institutional and logistical questions that attend to practices of gearing ideas towards key audiences. It's not just: How do I articulate a non-anthropocentric worldview? Rather, it is: How do I articulate a non-anthropocentric worldview in a way that can be implemented and realized within a given set of audiences? Curiosity *plus respicere* adds an extra-disciplinary dimension to research, because the intended audience is expanded beyond disciplinary peers. In a thin version, this could mean a kind of marketing plan to call attention to peer-reviewed publications once they are in print. But in thicker, richer versions, the expanded audience base would alter the very conduct of research – changing the standards of rigour, the language, and the media used.

This point can be seen as lurking within Callicott's own version of serendipity, which makes it different from Bush's disciplinary-based version. Callicott defines philosophy to include "theory in the natural and social sciences, as well as in history and the humanities, generally, not just philosophy as practiced by today's narrowly specialized professional philosophers." By this definition, Aldo Leopold, Albert Einstein, and Rachel Carson all count as philosophers. This expansion of the term makes a world of difference. Leopold and Carson, for example, found ways to philosophize within non-disciplinary contexts. It's not just that they worked in government. Both *A Sand County Almanac* and *Silent Spring* were written for an audience other than specialized academic peers. In other words, they were not relying on someone else to pull jargon-laden scholarship out of a peer-reviewed repository of knowledge and translate it for citizens and policymakers. Their pathway to impact is more direct and comprehensible.

Unlike Bush, Callicott is not primarily interested in offering a defence of scientific autonomy via self-governance of the disciplines. Rather, his interest is in making the argument that "what we do depends upon what we think." This leaves the specific type of dependency relationship between thought and action wide open. It could be that disciplinary research trickles down into action, but that is by no means the only way of doing philosophy and linking it up with impacts. You might also conjure up an idea and put it into a novel, a short video, a policy white paper, a community talk, an act of civil disobedience, etc. This plurality of research activities is essential to curiosity *plus respicere* – thinking not just about *what* ideas to ponder but also about *how* to present ideas and to *whom*. Indeed, we think the key, generally, for philosophy and the humanities to survive and flourish in the age of impact is to become pluralists. Not everyone has to engage in non-disciplinary research activities, but a solid number of humanists doing so would help provide herd immunity for everyone in an age of increased scrutiny. In short, it's time to break the radical monopoly of the disciplinary model of knowledge production.

CONCLUSION

In city planning the term 'complete street' means a street designed to accommodate automobiles as well as bicyclists and pedestrians. It's one where people on pedal and foot are no longer confined to the margins and forced to work according to the structural needs of automobiles. They have crosswalks, sidewalks, and bike lanes. Cars still use complete streets. It's just that others become equally recognized and legitimate participants in their own right. Analogously, 'complete philosophy' or humanities would still include disciplinary scholarship. Some people might devote their entire time to it. Others might do it minimally or not at all. But the key is that non-disciplinary models of research will gain recognition as the 'real deal'. We imagine a more balanced world of academic humanities where every department self-consciously establishes a certain ratio of disciplinary and non-disciplinary research as part of their identity. This will give them the chance to tell more robust and diversified stories about their pathways to impact.

In the spirit of pluralism, consider again Borgmann's distinction between holding the hand of the sick and inventing a cure for the disease. It's not just that it is a false either/or, as if one could not do both. It's that there are other roles to play. Someone, for example, has to collect data about the disease and someone has to test, administer, and monitor the cure. There is, in other words, intellectual activity to be done at the bedside and in other social spaces – *in media res* and at a micro- and meso-scale where fundamental discoveries struggle to take particular shape and where the particulars inform thought about the fundamentals. This is why NIH puts so many resources into translational or bench-to-bedside research.

Louis Pasteur blurred the boundaries between laboratory and world. His laboratory was as much out in the field with people and animals as it was back in a room with instruments. His were multi-dimensional and fluid relationships between knowledge and action. In recognition of this, Donald Stokes (1997) expanded the one-dimensional, linear model of serendipity into a quadrant of research types; Pasteur's quadrant is classified as "use-inspired basic research." The impact regime is providing philosophy with a chance to create its own Pasteur's quadrant. Thriving in this new accountability, culture will require institutionalizing something like use-inspired basic research in the humanities. The 'basic' element can still remain as we follow our curiosity, but it needs the *plus respicere* of being inspired and informed by the uses our curiosity might serve. Humanists, in short, will need to not only talk *about* matters of ultimate significance; they will also need to talk *with* the people involved when those ultimate matters are driving particular events. Midwifing a new worldview, not to mention simply helping people make sense of their lives, can happen at the bench and the bedside – or rather, in the armchair and in the fray.

Bibliography

Akers, Beth, and Chingos, Matthew M. 2014. "Is a Student Loan Crisis on the Horizon?" Brookings Institution. http://www.brookings.edu/~/media/research/files/reports/2014/06/24-student-loan-crisis-akers-chingos/is-a-student-loan-crisis-on-the-horizon.pdf.

Alcoff, Linda Martín. 2002. "Does the Public Intellectual Have Intellectual Integrity?" *Metaphilosophy*, vol. 33, no. 5, pp. 521–34.

Alexander, Shana. 1962. "They Decide Who Lives, Who Dies." *Life*, November 9, pp. 102–25.

Allan, Nicole, and Thompson, Derek. 2013. "The Myth of the Student-Loan Crisis." *The Atlantic*, March 2013, at http://www.theatlantic.com/magazine/archive/2013/03/myth-student-loan-crisis/309231/.

Andre, Judith. 2002. *Bioethics as Practice.* Chapel Hill, NC: The University of North Carolina Press.

Annas, George. 1993. *Standard of Care: The Law of American Bioethics.* New York: Oxford University Press.

American Philosophical Association. 1977. "Special Report: Sketches From Non-Academia: A Report of the APA Subcommittee on Non-Academic Careers for Philosophers." *Metaphilosophy*, vol. 8, no. 2–3, pp. 232–34.

Archard, David. 2009. "Applying Philosophy: A Response to O'Neill." *Journal of Applied Philosophy*, vol. 26, no. 3, pp. 239–44.

Archard, David, and Susan Mendus. 2009. "Introduction." *Journal of Applied Philosophy*, vol. 26, no. 3, pp. 217–18.

Aristotle. 1999. *Nicomachean Ethics.* Martin Ostwald, trans. Upper Saddle River, NJ: Prentice Hall.

Arnold, Matthew. 1869/2009. *Culture and Anarchy.* Oxford: Oxford University Press.

Arvan, Marcus. 2015. "Comment to 'Readers Identify the Most Important Issues in the Profession'." *Leiter Reports* Blog.

Baker, Robert. 2007. "A History of Codes of Ethics for Bioethicists." In Eckenwiler, Lisa A., and Felicia Cohn, eds. *The Ethics of Bioethics: Mapping the Moral Landscape.* Baltimore, MD: The Johns Hopkins University Press.

Baron, Jonathan. 2006. *Against Bioethics*. Cambridge, MA: MIT Press.

Bauerlein, Mark. 2011. *The Digital Divide: Arguments for and Against Facebook, Google, Texting, and the Age of Social Networking*. New York: Penguin.

Bauerlein, Mark, et al. 2010. "We Must Stop the Avalanche of Low-Quality Research." *Chronicle of Higher Education*, 13 June 2010.

Beauchamp, Tom, and James Childress. 1979. *Principles of Biomedical Ethics*, 1st ed. New York: Oxford University Press.

Beecher, Henry. 1966. "Ethics and Clinical Research." *New England Journal of Medicine*, vol. 274, pp. 1354–60.

Benneworth, Paul. 2015. "Putting Impact into Context: The Janus Face of the Public Value of Arts and Humanities Research." *Arts & Humanities in Higher Education*, vol. 14, no. 1, pp. 3–8.

Berlinger, Nancy. 2004. "Field Notes: From 'Idea' to 'Impact'." *Hastings Center Report*, vol. 34, no. 4.

Bérubé, Michael, and Jennifer Ruth. 2015. "The Humanities, Higher Education, and Academic Freedom: Three Necessary Arguments." *Times Higher Education*, April 9.

Biswas, Asit K., and Julian Kirchherr. 2015. "Prof, No One is Reading You." *The Straits Times*, April 11.

Boardman, Craig. 2014. "The New Visible Hand: Understanding Today's R&D Management." *Issues in Science and Technology*, vol. 2, pp. 23–26.

Bordogna, Francesca. 2008. *William James at the Boundaries*. Chicago: University of Chicago Press.

Borgmann, Albert. 1995. "Does Philosophy Matter?" *Technology in Society*, vol. 17, no. 3, pp. 295–309.

Bowie, Norman. 1982. "Applied Philosophy – Its Meaning and Justification." *International Journal of Applied Philosophy*, vol. 1, no. 1, pp. 1–18.

Breck, John, and Lyn Breck. 2005. *Stages on Life's Way: Orthodox Thinking on Bioethics*. Crestwood, NY: St. Vladimir's Seminary Press.

Briggle, Adam. 2005. "Visions of Nantucket." *Environmental Philosophy*, vol. 2, no. 1, pp. 54–67.

Briggle, Adam. 2010. *A Rich Bioethics: Public Policy, Biotechnology, and the Kass Council*. Notre Dame, IN: University of Notre Dame Press.

Briggle, Adam. 2014. "Opening the Black Box: The Social Outcomes of Scientific Research." *Social Epistemology*, vol. 28, no. 2, pp. 153–66.

Briggle, Adam. 2015. *A Field Philosopher's Guide to Fracking*. New York: Liveright.

Briggle, Adam, and Robert Frodeman. 2011. "Creating a 21st Century Philosophy." *The Chronicle Review*, December 18.

Briggle, Adam, Robert Frodeman, and Kelli Barr. 2015. "Achieving Escape Velocity: Breaking Free from the Impact Failure of Applied Philosophy." London School of Economics, *Impact of Social Sciences Blog*, April 27.

Brister, Evelyn. 2016. "Disciplinary Capture and Epistemological Obstacles to Interdisciplinary Research: Lessons from Central African Conservation Disputes." *Studies in History and Philosophy of Biological and Biomedical Sciences*.

Brooks, David. 2013. "The Humanist Vocation." *New York Times*, June 20.

Bruner, Michael, and Max Oelschlager. 1994. "Rhetoric, Environmentalism, and Environmental Ethics." *Environmental Ethics*, vol. 16, no. 4, pp. 377–96.

Bush, Vannevar. 1945. "Science – the Endless Frontier." Washington, DC: US Government Printing Office.

Callahan, Daniel. 1973. "Bioethics as a Discipline." *Hastings Center Studies*, vol. 1, pp. 66–73.

Callahan, Daniel. 1998. *False Hopes: Why America's Quest for Perfect Health is a Recipe for Failure*. New York: Simon and Schuster.

Callahan, Daniel. 2003. "Individual Good and Common Good: A Communitarian Approach to Bioethics." *Perspectives in Biology and Medicine*, vol. 46, no. 4, pp. 496–507.

Callahan, Daniel. 2004. "Bioethics." In *The Encyclopedia of Bioethics*, 3rd ed., edited by Stephen Post, pp. 278–87. New York: Macmillan.

Callahan, Daniel. 2005. "Bioethics and the Culture Wars." *Cambridge Quarterly of Healthcare Ethics*, vol. 14, no. 4, pp. 424–31.

Callicott, Baird. 1999. "Environmental Philosophy *Is* Environmental Activism: The Most Radical and Effective Kind." In *Beyond the Land Ethic: More Essays in Environmental Philosophy*, pp. 27–44. Albany, NY: State University of New York Press.

Carey, Kevin, 2015. "Student Debt is Worse Than You Think." *New York Times*, October 7.

Carnap, Rudolph. 1963. *The Philosophy of Rudolf Carnap*. P.A. Schilp, ed. La Salle, IL: Open Court.

Carson, Andrew. 2013. "Graduate School Philosophy Placement Records in the US and CA: Will I Get a Job?" *Philosophy News*, October.

Carson, Rachel. 1963. *Silent Spring*. Boston, MA: Houghton Mifflin.

Casey, Michael, and Robert Hackett. 2014. "The 10 Biggest R&D Spenders Worldwide." *Fortune*, November 17.

Chambers, Tod. 1999. *The Fiction of Bioethics: Cases as Literary Texts*. New York: Routledge.

Charo, R. Alta. 2004. "Passing on the Right: Conservative Bioethics is Closer than it Appears." *Journal of Law, Medicine, and Ethics*, vol. 32, no. 2, pp. 307–14.

Childress, James. 1970. "Who Shall Live When Not All Can Live?" *Soundings*, vol. 53, no. 4, pp. 339–55.

Clement, Grace. 1996. *Care, Autonomy, and Justice: Feminism and the Ethic of Care*. Boulder, CO: Westview Press.

Colberg, Magda. 1986. "The Application of Logic to Psychometrics." *International Journal of Applied Philosophy*, vol. 3. no. 1, pp. 59–64.

Collins, H. M. and R. J. Evans. 2002. "The Third Wave of Science Studies: Studies of Expertise and Experience." *Social Studies of Sciences*, vol. 32, no. 2, pp. 235–96.

Collins, Randall, 1998. *The Sociology of Philosophies: A Global Theory of Intellectual Change*. Cambridge, MA: Harvard University Press.

Cooper, Mark G. and John Marx. 2014. "Crisis, Crisis, Crisis: Big Media and the Humanities Workforce." *Differences: A Journal of Feminist Cultural Studies*, vol. 24, no. 3, pp. 127–59.

Crow, Michael M., and William B. Dabar. 2015. *Designing the New American University*. Baltimore, MD: Johns Hopkins University Press.

Darwin, Charles. 1959. *The Origin of Species*. New York: Bantam Classics.

deGrasse Tyson, Neil. 2014. "Neil deGrasse Tyson Returns Again." *Nerdist Podcast*, March 7. http://nerdist.com/nerdist-podcast-neil-degrasse-tyson-returns-again/.

Delbanco, Andrew. 2015. "Our Universities: The Outrageous Reality." *New York Review of Books*, July 9.

Deleuze, Gilles, and Felix Guattari. 1980/1987. *A Thousand Plateaus: Capitalism and Schizophrenia*. Minneapolis, MN: University of Minnesota Press.

Deresiewicz, William. 2015. Excellent Sheep: The Miseducation of the American Elite and the Way to a Meaningful Life. New York: Free Press.

Dewey, John. 1917. "The Need for a Recovery of Philosophy." In *Creative Intelligence: Essays in the Pragmatic Attitude*, pp. 3–69. New York: Henry Holt.

Dolhenty, Jonathan. 2008. "What do We Mean by Applied Philosophy?" *The Moral Liberal*. http://www.themoralliberal.com/2014/12/06/what-do-we-mean-by-applied-philosophy/.

Donoghue, Frank. 2008. *The Last Professors: The Corporate University and the Fate of the Humanities*. Bronx, NY: Fordham University Press.

Dorfman, Jeffrey. 2014. "Surprise: Humanities Degrees Provide Great Return on Investment." *Forbes*, November 20.

Douglas, Heather. 2016. "A History of the PSA Before 1970." Philosophy of Science Association website. Available at: http://www.philsci.org/about-the-psa/history-of-the-association.

Dyson, Freeman. 2012. "What Can You Really Know?" *New York Review of Books*, November 8.

Dzur, Albert, and Daniel Levin. 2004. "The 'Nation's Conscience:' Assessing Bioethics Commissions as Public Forums." *The Kennedy Institute of Ethics Journal*, vol. 14, no. 4, pp. 333–60.

Eckenwiler, Lisa A., and Felicia Cohn, eds. 2007. *The Ethics of Bioethics: Mapping the Moral Landscape*. Baltimore, MD: The Johns Hopkins University Press.

Engelhardt, Tristram Jr. 1996. *The Foundations of Bioethics*, 2nd ed. New York: Oxford University Press.

Evans, John. 2011. *The History and Future of Bioethics: A Sociological View*. New York: Oxford University Press.

Fehr, Carla, and Kathryn S. Plaisance. 2010. "Socially Relevant Philosophy of Science: An Introduction." *Synthese*, vol. 177, no. 3, pp. 301–16.

Fish, Stanley. 2008. "Will the Humanities Save Us?" *The New York Times*, January 6, 2008.

Fisher, Erik, and R. L. Mahajan. 2006. "Midstream Modulation of Nanotechnology Research in an Academic Laboratory." *American Society of Mechanical Engineers International Mechanical Engineering Congress and Exposition*, Chicago.

Fleetwood, Janet. 1987. "Medical Ethics in the Clinical Setting: Challenging the M. D. Monopoly." *International Journal of Applied Philosophy*, vol. 3, no. 4, pp. 61–68.

Fletcher, John C., and Miller, Franklin G. 1996. The Promise and Perils of Public Bioethics. *The Ethics of Research Involving Human Subjects*, ed. Harold Y. Vanderpool, pp. 155–84. Frederick, MD: University Publishing Group.

Frodeman, Robert. 1995. "Geological Reasoning: Geology as an Interpretive and Historical Science." *Geological Society of America Bulletin*, vol. 107, no. 8, pp. 960–68.

Frodeman, Robert. 2006. "The Policy Turn in Environmental Philosophy." *Environmental Ethics*, vol. 28, no. 1, pp. 3–20.

Frodeman, Robert. 2007a. "The Role of Humanities Policy in Public Science." In *Public Science in Liberal Democracies*, pp. 111–20. Toronto, CA: University of Toronto Press.

Frodeman, Robert. 2007b. "The Future of Environmental Philosophy." *Ethics & the Environment*, vol. 12, no. 2: 120–22.

Frodeman, Robert. 2008. "Philosophy Unbound: Environmental Thinking at the End of the Earth." *Environmental Ethics*, vol. 30, no. 3, pp. 47–61.

Frodeman, Robert. 2009. "Intellectual Merit and Broader Impact: The National Science Foundation's Broader Impacts Criterion and the Question of Peer Review." *Social Epistemology*, vol. 23, no. 3–4, pp. 337–45.

Frodeman, Robert. 2012. "Philosophy Dedisciplined." *Synthese*, vol. 190, no. 11, pp. 1917–36.

Frodeman, Robert. 2013. *Sustainable Knowledge: A Theory of Interdisciplinarity*. Basingstoke: MacMillan/Palgrave Press.

Frodeman, Robert, and Carl Mitcham. 2004. "New Directions in the Philosophy of Science: Toward a Philosophy of Science Policy." *Philosophy Today*, vol. 48, no. 5, pp. 3–15.

Frodeman, Robert, and Adam Briggle. 2012. "The Dedisciplining of Peer Review." *Minerva*, vol. 5, no. 1, pp. 3–19.

Fuller, Steve, and Veronica Lipinska. 2014. *The Proactionary Imperative: A Foundation for Transhumanism*. London: Palgrave MacMillan.

Fuller, Steve. 2011. *Humanity 2.0*. London: Palgrave Macmillan.

Fullinwider, Robert K. 1989. "Against Theory, Or: Applied Philosophy – A Cautionary Tale." *Metaphilosophy*, vol. 20, no. 3–4, pp. 222–34.

GAO. 2014. "Higher Education State Funding Trends and Policies on Affordability." *US Government Accountability Office*, GAO-15-151, December.

Gardiner, Stephen. 2007. "Environmental Midwifery and the Need for an Ethics of the Transition: A Quick Riff on the Future of Environmental Ethics." *Ethics & the Environment*, vol. 12, no. 2, pp. 122–23.

Gibbons, M., C. Limoges, H. Nowotny, S. Schwartzman, P. Scott, and M. Trow. 1994. *The New Production of Knowledge: The Dynamics of Science and Research in Contemporary Societies*. London: Sage.

Girill, T. R. 1984. "Philosophy's Relevance to Technical Writing." *International Journal of Applied Philosophy*, vol. 2, no. 2, pp. 89–95.

Goldstein, Rebecca. 2014. *Plato at the Googleplex: Why Philosophy Won't Go Away*. New York: Pantheon.

Graff, Gerald. 2007. *Professing Literature: An Institutional History*. Chicago: University of Chicago Press.

Grafton, Anthony. 2009. "A Sketch Map of a Lost Continent: The Republic of Letters." *Republics of Letters: A Journal for the Study of Knowledge, Politics, and the Arts*, vol. 1, no. 1, pp. 1–18.

Guston, David. 2000. *Between Politics and Science: Assuring the Integrity and Productivity of Research*. Cambridge: Cambridge University Press.

Hale, Ben. 2011. "The Methods of Applied Philosophy and the Tools of the Policy Sciences." *International Journal of Applied Philosophy*, vol. 25, no. 2, pp. 215–32.

Harden, Nathan. 2012. "The End of the University as We Know It." *The American Interest*, vol. 8, no. 3.

Hargrove, Eugene. 1979. "The State of the Journal." *Environmental Ethics*, vol. 1, no. 1, pp. 291–92.

Hargrove, Eugene. 1989. "The Future is Now." *Environmental Ethics*, vol. 11, no. 4, pp. 291–92.

Hargrove, Eugene. 1998. "After Twenty Years." *Environmental Ethics*, vol. 20, no. 4, pp. 339–40.

Hargrove, Eugene. 2000. "The Next Century and Beyond." *Environmental Ethics*, vol. 22, no. 1, p. 3.

Hargrove, Eugene. 2003. "What's Wrong? Who's to Blame?" *Environmental Ethics*, vol. 25, no. 1, pp. 3–4.

Hargrove, Eugene. 2012. "Biology, Environmental Ethics, and Policy." *Environmental Ethics*, vol. 34, no. 1, pp. 3–4.

Harpham, Geoffrey G. 2011a. *The Humanities and the Dream of America*. Chicago: University of Chicago Press.

Harpham, Geoffrey G. 2011b. "Why We Need the 16,772nd Book on Shakespeare." *Qui Parle*, vol. 20, no. 1, pp. 109–16.

Hawking, Steven, and Leonard Mlodinow. 2010. *The Grand Design*. New York: Bantam Books.

Heidegger, Martin. 1993. "The End of Philosophy and the Task of Thinking." In *Basic Writings*, 2nd ed., edited by David Farrell Krell, pp. 427–49. New York: HarperCollins.

Hicks, Diana, Paul Wouters, Ludo Waltman, Sarah de Rijcke, and Ismael Rafols. 2015. "Bibliometrics: The Leiden Manifesto for Research Metrics." *Nature*, vol. 520, no. 7548, pp. 429–31.

Hoeller, Keith. 2014. "The Wal-Mart-ization of Higher Education: How Young Professors are Getting Screwed." *Salon*, February 16.

Holbrook, J. Britt, ed. 2009. Special Issue: US National Science Foundation's Broader Impacts Criterion. *Social Epistemology*, vol. 23, no. 3–4, pp. 177–405.

Illich, Ivan. 1973. *Tools for Conviviality*. London: Marion Boyars.

Jacoby, Russell. 2000. *The Last Intellectuals American Culture in the Age of Academe*. New York: Basic Books.

Jasanoff, Sheila. 2010. "A Field of its Own: The Emergence of Science and Technology Studies." In *Oxford Handbook of Interdisciplinarity*, edited by Robert Frodeman, Julie-Thompson Klein, and Carl Mitcham, pp. 191–205. Oxford: Oxford University Press.

Jaspers, Karl. 1963. *The Atom Bomb and the Future of Man.* Chicago: University of Chicago Press.

Johnston, Josephine. 2008. "Field Notes: Talking Points." *Hastings Center Report,* vol. 38, no. 3.

Jonsen, Albert. 1998. *The Birth of Bioethics.* New York: Oxford University Press.

Joy, Bill. 2000. "Why the Future Does Not Need Us." *Wired Magazine,* April.

Juengst, Eric T. 1996. "Self-Critical Federal Science? The Ethics Experiment Within the U.S. Human Genome Project." *Social Philosophy and Policy,* vol. 13, no. 2, pp. 63–95.

Kagan, Connie. 1985. "Philosophy and Animal Protection Legislation: A Personal Account." *International Journal of Applied Philosophy,* vol. 2, no. 4, pp. 95–99.

Kasachkoff, Tziporah. 1992. "Some Complaints About and Some Defenses of Applied Philosophy." *International Journal of Applied Philosophy,* vol. 7, no. 1, pp. 5–9.

Kass, Leon. 2002. *Life, Liberty and the Defense of Dignity.* San Francisco: Encounter Books.

Katz, Eric, and Lauren Oechsli. 1993. "Moving beyond Anthropocentrism: Environmental Ethics, Development, and the Amazon." *Environmental Ethics,* vol. 15, no. 1, pp. 49–59.

Katzner, Louis. 1982. "Applied Philosophy and the Role of the Philosopher." *International Journal of Applied Philosophy,* vol. 1, no. 2, pp. 31–39.

Kelbessa, Workineh. 2004. "Can African Environmental Ethics Contribute to Environmental Policy in Africa?" *Environmental Ethics,* vol. 36, no. 1, pp. 31–61.

Kiley, Kevin. 2013. "Another Liberal Arts Critic." *Inside Higher Ed,* January 30.

Kitcher, Philip. 2011. "Philosophy Inside Out." *Metaphilosophy,* vol. 42, no. 3, pp. 248–60.

Knobe, Joshua, and Shaun Nichols, eds. 2008. *Experimental Philosophy.* Oxford: Oxford University Press.

Kon, A. A. 2009. "The Role of Empirical Research in Bioethics." *American Journal of Bioethics,* vol. 9, nos. 6–7, pp. 59–65.

Kuklick, Bruce. 1978. *The Rise of American Philosophy.* New Haven, CT: Yale University Press.

Kuklick, Bruce. 2001. *A History of Philosophy in America, 1720-2000.* Oxford: Oxford University Press.

Lam, Bourree. 2015. "The Earning Power of Philosophy Majors." *The Atlantic,* September 3. http://www.theatlantic.com/notes/2015/09/philosophy-majors-out-earn-other-humanities/403555/.

Latour, Bruno. 1993. *We Have Never Been Modern.* Cambridge, MA: Harvard University Press.

Leiter, Brian, ed. 2007. *The Future for Philosophy.* Oxford: Oxford University Press.

Leiter, Brian. 2010. "What is the NY Times Thinking?" *Leiter Reports: A Philosophy Blog,* May 16. http://leiterreports.typepad.com/blog/2010/05/what-is-the-ny-times-thinking.html.

Leopold, Aldo. 1949. *A Sand County Almanac.* Oxford: Oxford University Press.

Levine, Carol. 2007. "Analyzing Pandora's Box: The History of Bioethics." In Eckenwiler, Lisa A., and Felicia Cohn, eds. *The Ethics of Bioethics: Mapping the Moral Landscape.* Baltimore, MD: The Johns Hopkins University Press.

Light, Andrew, and Eric Katz. 1996. *Environmental Pragmatism.* New York: Routledge.

Loncar, Samuel. 2016. "Why Listen to Philosophers?" *Metaphilosophy*, vol. 16, no. 1, pp. 3–25.

Loux, Michael J., and Dean W. Zimmerman. 2003. *The Oxford Handbook of Metaphysics.* Oxford: Oxford University Press.

MacIntyre, Alasdair. 1984. *After Virtue*, 2nd ed. Notre Dame: University of Notre Dame Press.

Mandler, Peter. 2015. "Rise of the Humanities." Aeon, December 17. https://aeon.co/essays/the-humanities-are-booming-only-the-professors-can-t-see-it.

Manes, Christopher. 1988. "Philosophy and the Environmental Task." *Environmental Ethics*, vol. 10, no. 1, pp. 75–82.

Mathis-Lilley, Ben. 2015. "Frank Gifford Had CTE, his Family Says, as Doctors Call for End of High School Football." *Slate*, November 25.

McCumber, John. 2001. *Time in the Ditch: American Philosophy and the McCarthy Era.* Evanston, IL: Northwestern University Press.

McMahan, Jeff. 2009. "Five Questions about Normative Ethics." In T. S. Petersen and J. Ryberg, eds. *Normative Ethics: 5 Questions.* New York: Automatic Press.

Meagher, Sharon. 2013. "Public Philosophy: Revitalizing Philosophy as a Civic Discipline." Report to the Kettering Foundation, January 13.

Meagher, Sharon, and Ellen K. Feder. 2010. "The Troubled History of Philosophy and Deliberative Democracy." *Journal of Public Deliberation*, vol. 6, no. 1, article 6.

Minteer, Ben A. 2007. "The Future of Environmental Philosophy." *Ethics & the Environment*, vol. 12, no. 2, pp. 132–33.

Mitcham, Carl. 1994. "Engineering Design Research and Social Responsibility." In Kristin Shrader-Frechette, ed. *Ethics of Scientific Research*, pp. 153–55. Lanham, MD: Rowman and Littlefield.

MLA. 2014. *Report of the Task Force on Doctoral Study in Modern Language and Literature.* Accessed 20 October 2015 at: https://www.mla.org/report_doctoral_study_2014.

Moreno, Jonathan. 2005. "The End of the Great Bioethics Compromise." *Hastings Center Report*, vol. 35, no. 1, pp. 14–15.

Moreno, Jonathan D. 2010. *Progress in Bioethics: Science, Policy, and Politics.* Cambridge, MA: MIT Press.

Morrison, Donald R., ed. 2011. *The Cambridge Companion to Socrates.* Cambridge: Cambridge University Press.

National Bioethics Advisory Commission. 1997. *Cloning Human Beings.* Bethesda, MD: U.S. Government Printing Office.

National Commission for the Protection of Human Subjects of Biomedical and Behavioral Research. 1979. *The Belmont Report: Ethical Principles and Guidelines for the Protection of Human Subjects of Research.* Bethesda, MD: The Commission.

Newfield, Christopher. 2011. *Unmaking the Public University: The Forty-Year Assault on the Middle Class.* Cambridge, MA: Harvard University Press.

Nietzsche, Friedrich. 1886/1989. *Beyond Good and Evil.* Walter Kaufman, trans. New York: Vintage.

Norton, Bryan G. 1994. *Toward Unity among Environmentalists*. Oxford: Oxford University Press.

Norton, Brian. 2007. "The Past and Future of Environmental Philosophy." *Ethics & the Environment*, vol. 12, no. 2, pp. 134–36.

Nussbaum, Martha. 2010. *Not for Profit: Why Democracy Needs the Humanities*. Princeton, NJ: Princeton University Press.

O'Malley, Maureen, Jane Calvert, and John Dupré. 2007. "The Study of Socioethical Issues in Systems Biology." *American Journal of Bioethics*, vol. 7 no. 4, pp. 67–78.

O'Neill, Onora. 2009. "Applied Ethics: Naturalism, Normativity and Public Policy." *Journal of Applied Philosophy*, vol. 26, no. 3, pp. 219–30.

O'Rourke, Michael, and Steve Crowley. 2012. "Philosophical Intervention and Cross-disciplinary Science: The Story of the Toolbox Project." *Synthese*, vol. 190, no. 11, pp. 1937–54.

Passmore, John. 1974. *Man's Responsibility for Nature: Ecological Problems and Western Traditions*. New York: Charles Scribner's Sons.

Pedersen, Arthur Paul. 2012. "Two Reasons for Abolishing the PGR." *Choice and Inference*, February 11. Accessed February 11, 2016, at: http://choiceandinference.com/2012/04/24/two-reasons-for-abolishing-the-pgr/.

Pellegrino, Edmund D. 2006. "Toward a Reconstruction of Medical Morality." *American Journal of Bioethics*, vol. 6, no. 2, pp. 65–71.

Pielke, Jr., Roger A. and Radford Byerly, Jr. 1998. "Beyond Basic and Applied." *Physics Today*, vol. 51, no. 2, pp. 42–46.

Polanyi, Michael. 1962. "The Republic of Science." *Minerva*, vol. 1, no. 1, pp. 54–74.

Potter, V. R. 1970. "Bioethics: The Science of Survival." *Perspectives in Biology and Medicine*, vol. 14, pp. 127–53.

Potter, Van Renesselar. 1971. *Bioethics: Bridge to the Future*. Englewood Cliffs, NJ: Prentice-Hall.

Priestley, Joseph. 1775. *The History and Present State of Electricity* (2 vols.), 3rd ed. London.

Puka, Bill. 1986. "Applied Philosophy–Taking a Stand." *International Journal of Applied Philosophy*, vol. 3, no. 1, pp. 69–84.

Pulizzi, James. 2014. "In the Near Future, Only Very Wealthy Colleges Will Have English Departments." *The New Republic*, June 8. Accessed February 7, 2016, at: https://newrepublic.com/article/118025/advent-digital-humanities-will-make-english-departments-pointless.

Quine, Willard V. O. 1981. *Theories and Things*. Cambridge, MA: Belknap Press of Harvard University Press.

Ramsey, Paul. 1976. "Prolonged Dying: Not Medically Indicated." *Hastings Center Report*, vol. 6, no. 1, pp. 14–17.

Reisch, George A. 2005. *How the Cold War Transformed Philosophy of Science*. Cambridge: Cambridge University Press.

Rinella, Michael A. 2011. *Pharmakon: Plato, Drug Culture, and Identity in Ancient Athens*. Lanham, MD: Lexington Books.

Robert, Jason S. 2007. "Systems Bioethics." *American Journal of Bioethics*, vol. 7, no. 4, pp. 80–82.

Romano, Carlin. 2012. *America the Philosophical*. New York: Vintage.

Rorty, Richard. 1979. *Philosophy and the Mirror of Nature.* Princeton, NJ: Princeton University Press.

Sarewitz, Dan. 1996. *Frontiers of Illusion.* Philadelphia, PA: Temple University Press.

Sathian, Sanjena. 2016. "The 21st Century Philosophers." *OZY,* January 4.

Schmidt, Ben 2013. "A Crisis in the Humanities?" *Chronicle of Higher Education,* June 10, 2013.

SciSIP, 2015. SciSIP Call for Proposals. *National Science Foundation.* http://www.nsf.gov/funding/pgm_summ.jsp?pims_id=501084.

Shapin, Steven. 2008. *The Scientific Life: A Moral History of a Late Modern Vocation.* Chicago: University of Chicago Press.

Shattuck, Roger. 1997. *Forbidden Knowledge: From Prometheus to Pornography.* New York: St. Martin's Press.

Sheehan, Mark, and Michael Dunn. 2012. "On the Nature and Sociology of Bioethics." *Health Care Analysis,* vol. 21, pp. 54–69.

Sinclair, Upton. 1923. *The Goose-Step: A Study of American Education.* Available online at a number of locations.

Small, Helen. 2013. *The Value of the Humanities.* Oxford: Oxford University Press.

Smith, Adam. 1776/2003. *The Wealth of Nations.* New York: Bantam Classics.

Snowden, Fraser. 1982. "Bringing Philosophy into the Hospital: Notes of a Philosopher-in-Residence." *International Journal of Applied Philosophy,* vol. 1, no. 3, pp. 67–81.

Soames, Scott. 2005. "How Many Grains Make a Heap?" *London Review of Books, Letters,* vol. 27, no. 5, March 3.

Stanley, Jason. 2010. "The Crisis of Philosophy." *Inside Higher Education,* April 5.

Stevenson, Leslie. 1970. "Applied Philosophy." *Metaphilosophy,* vol. 1, no. 3, pp. 258–67.

Stokes, Donald. 1997. *Pasteur's Quadrant: Basic Science and Technological Innovation.* Washington, DC: Brookings Institution Press.

Strauss, Leo. 1952. *Persecution and the Art of Writing.* New York: The Free Press.

Strauss, Leo. 1958. *Natural Right and History.* Chicago: University of Chicago Press.

Taylor, Devin, and Adam Briggle. 2015. "Time to Ride Wave of Renewable Energy." *Denton Record Chronicle,* November 14. http://www.dentonrc.com/opinion/columns-headlines/20151114-time-to-ride-wave-of-renewable-energy.ece.

Thompson, Paul B. 2010. *The Agrarian Vision: Sustainability and Environmental Ethics.* Lexington, KY: University Press of Kentucky.

Thompson, Rebecca. 1990. "A Refutation of Environmental Ethics." *Environmental Ethics,* vol. 12, no. 2, pp. 147–60.

Thurgood, L., M. J. Golladay, and S. T. Hill. 2006. *U.S. Doctorates in the 20th Century: Special Report.* National Science Foundation.

Torgerson, Jon N. 1977. "The Philosopher-in-Residence: An Approach to Teaching Philosophy." *Metaphilosophy,* vol. 8, no. 2–3, pp. 215–21.

Toulmin, Stephen. 1982. "How Medicine Saved the Life of Ethics." *Perspectives in Biology and Medicine,* vol. 2, no. 4, pp. 736–50.

Turner, Leigh. 2009. "Does Bioethics Exist?" *Journal of Medical Ethics*, vol. 35, pp. 778–80. doi:10.1136/jme.2008.028605.

Unger, Peter. 2014. *Empty Ideas: A Critique of Analytic Philosophy*. Oxford: Oxford University Press.

United States Office of Technology Assessment. 1993. *Biomedical Ethics in U.S. Public Policy*. OTA-PB-BBS-105. Washington, DC: U.S. Government Printing Office.

Weiss, Mitchell. 2015. "The Way We Look at Student Loan Debt is Dangerously Wrong." *Fox Business*, February 25.

Wellmon, Chad. 2015. *Organizing Enlightenment: Information Overload and the Invention of the Modern Research University*. Baltimore, MD: Johns Hopkins University Press.

Wittkower, D. E., Evan Selinger, and Lucinda Rush. 2013. "Public Philosophy of Technology: Motivations, Barriers, and Reforms." *Techne: Research in Philosophy and Technology*, vol. 17, no. 2, pp. 179–200.

Young, Christopher. 2004. "How to Teach Introduction to Applied Ethics." Available at http://www.chrisyoung.net/prose/essays/how-to-teach-introduction-to-applied-ethics/.

Zakaria, Fareed. 2015. *In Defense of a Liberal Education*. New York: W.W. Norton.

Index

abortion, 94, 106
academic:
    culture, 36, 81, 131;
    freedom, 39, 119;
    knowledge, x, 8;
    metrics, 23, 36;
    politics, 45;
    research, 4, 21, 23
Académie Française, 60
activism, 24, 106
aesthetics, 8, 13, 77, 118
agricultural ethics, 84
American Philosophical Association
    (APA), 36,40, 45, 69–70, 128,
    130–31
analytic philosophy, x, xi, 7, 19, 20, 35,
    50, 66
anarchy, 141
Andre, Judith, 101, 103
Aristotle, 55, 57, 60, 61, 78, 94,
    116, 143
Arizona State University (ASU),
    131–32
authoritative knowledge, 61, 121

Belmont Report, 97, 105
Benedict, Ruth, 50
bibliometrics, 25, 110, 126, 135
Bildung, 64, 141

biology, 30, 37, 41, 61, 86, 97, 109
Boon, Mieke, 129
Borgmann, Albert, 35, 145–46, 150
Bostrom, Nick, 122, 136
boundaries, 26, 61, 62, 94, 122
boundary work, 68, 84–85, 98–99
Breakthrough Institute, 122
broader impacts, 5, 25, 58–59, 71, 75n7,
    87–88, 103, 111–12, 129, 135,
    137, 139
Brooks, David, 38
Bruner, Michael, 86
Bush, Vannevar, 24–25, 75n3, 137–39,
    142, 144–45, 147, 149

Callahan, Daniel, 96, 104, 106, 133
Callicott, Baird, 72, 82, 146–49
capitalism, 21
Carson, Rachel, 149
Center for Applied Rationality, 122
climate science, 2, 49, 82, 88, 119
cognitivism, 82
Cold War, 21, 24, 65, 138, 143
commission paradigm, 74
complete streets, 149–50
contemporary philosophy, 1–2, 7, 9, 20,
    23, 30, 60, 65
continental philosophy, 7, 16, 20, 75
Cornell University, 64, 121

About the Author

Robert Frodeman is Professor of Philosophy in the Department of Philosophy and Religion Studies at the University of North Texas. He conducts research in environmental philosophy, science and technology policy, and questions concerning interdisciplinarity. He is the author of *Sustainable Knowledge: A Theory of Interdisciplinarity* (2013), and editor in chief of the *Oxford Handbook of Interdisciplinarity* (2016).

Adam Briggle is Associate Professor in the Department of Philosophy and Religion Studies at the University of North Texas. His teaching and research focus on the intersection of ethics, policy, science, and technology. He is the author of *A Rich Bioethics* (2010) and *A Field Philosopher's Guide to Fracking* (2015).